Dynamics of Topologically Generic Homeomorphisms

MEMOIRS
of the
American Mathematical Society

Number 783

Dynamics of Topologically Generic Homeomorphisms

Ethan Akin
Mike Hurley
Judy A. Kennedy

July 2003 • Volume 164 • Number 783 (end of volume) • ISSN 0065-9266

American Mathematical Society
Providence, Rhode Island

2000 *Mathematics Subject Classification.*
Primary 37B99.

Library of Congress Cataloging-in-Publication Data

Akin, Ethan, 1946–
Dynamics of topologically generic homeomorphisms / Ethan Akin, Mike Hurley, Judy A. Kennedy.
p. cm. — (Memoirs of the American Mathematical Society, ISSN 0065-9266 ; no. 783)
"Volume 164, number 783 (end of volume)."
Includes bibliographical references and index.
ISBN 0-8218-3338-3 (alk. paper)
1. Homeomorphisms. 2. Topological dynamics. 3. Metric spaces. I. Hurley, Mike, 1953– II. Kennedy, Judy A. (Judy Anita), 1947– III. Title. IV. Series.

QA3.A57 no. 783
[QA614]
510 s–dc21
[514] 2003048026

Memoirs of the American Mathematical Society

This journal is devoted entirely to research in pure and applied mathematics.

Subscription information. The 2003 subscription begins with volume 161 and consists of six mailings, each containing one or more numbers. Subscription prices for 2003 are $555 list, $444 institutional member. A late charge of 10% of the subscription price will be imposed on orders received from nonmembers after January 1 of the subscription year. Subscribers outside the United States and India must pay a postage surcharge of $31; subscribers in India must pay a postage surcharge of $43. Expedited delivery to destinations in North America $35; elsewhere $130. Each number may be ordered separately; *please specify number* when ordering an individual number. For prices and titles of recently released numbers, see the New Publications sections of the *Notices of the American Mathematical Society.*

Back number information. For back issues see the *AMS Catalog of Publications.*

Subscriptions and orders should be addressed to the American Mathematical Society, P. O. Box 845904, Boston, MA 02284-5904, USA. *All orders must be accompanied by payment.* Other correspondence should be addressed to 201 Charles Street, Providence, RI 02904-2294, USA.

Memoirs of the American Mathematical Society is published bimonthly (each volume consisting usually of more than one number) by the American Mathematical Society at 201 Charles Street, Providence, RI 02904-2294, USA. Periodicals postage paid at Providence, RI. Postmaster: Send address changes to Memoirs, American Mathematical Society, 201 Charles Street, Providence, RI 02904-2294, USA.

This publication is indexed in *Science Citation Index*®, *SciSearch*®, *Research Alert*®, *CompuMath Citation Index*®, *Current Contents*®*/Physical, Chemical & Earth Sciences.*
Printed in the United States of America.

♾ The paper used in this book is acid-free and falls within the guidelines established to ensure permanence and durability.
Visit the AMS home page at `http://www.ams.org/`

10 9 8 7 6 5 4 3 2 1 08 07 06 05 04 03

Contents

Abstract

The goal of this work is to describe the dynamics of generic homeomorphisms of certain compact metric spaces X. Here "generic" is used in the topological sense – a property of homeomorphisms on X is generic if the set of homeomorphisms with the property contains a residual subset (in the sense of Baire category) of the space of all homeomorphisms on X. The spaces X we consider are those with enough local homogeneity to allow certain localized perturbations of homeomorphisms; for example, any compact manifold is such a space. We show that the dynamics of a generic homeomorphism is quite complicated, with a number of distinct dynamical behaviors coexisting (some resemble subshifts of finite type, others, which we call 'generalized adding machines', appear strictly periodic when viewed to any finite precision, but are not actually periodic). Such a homeomorphism has infinitely many, intricately nested attractors and repellors, and uncountably many distinct dynamically-connected components of the chain recurrent set. We single out several types of these "chain components", and show that each type occurs densely (in an appropriate sense) in the chain recurrent set. We also identify one type that occurs generically in the chain recurrent set. We also show that, at least for X a manifold, the chain recurrent set of a generic homeomorphism is a Cantor set, so its complement is open and dense. Somewhat surprisingly, there is a residual subset of X consisting of points whose limit sets are chain components of a type other than the type of chain components that are residual in the space of all chain components. In fact, for each generic homeomorphism on X there is a residual subset of points of X satisfying a stability condition stronger than Lyapunov stability.

Received by the editor May 15, 2000, and in revised form February 12, 2002.

2000 *Mathematics Subject Classification.* Primary 37B99.

Introduction

0.1. Overview

It can be argued that in mathematics, there are no interesting generalities (self-referentially implying that this particular generality is uninteresting). The reason behind such an assertion is that to avoid counterexamples, one must either be quite specific, or else deal only in rather obvious tautologies.

In particular, even in trying to deal with a fairly specific question – that of describing dynamics of homeomorphisms on compact metric spaces – almost the first thing that one realizes is that there is little that one can say. There may be periodic points, or there may not; there may be dense orbits, or not; there may be relatively small parts of the space (to be called "attractors") whose dynamics control the long term dynamics of much larger parts of the space, or there may not.

There is a long tradition in dynamical systems of trying to surmount this difficulty by attempting to identify features of dynamical systems that are "typical", in one sense or another. (Unfortunately, the general principle that there are no interesting generalities includes the principle that there is no universally applicable definition of "typical".) After one settles upon a definition of "typical", one hopes to find properties that are common to all systems in a typical class. Often these properties are themselves properties that are shared by "typical" points of the phase space – thus allowing the possibility of using the word 'typical' in two different ways – typicality for the system and typicality of an initial condition.

In this paper we take what may be the simplest reasonable definitions of "typical" – we work strictly topologically, and so for us, "typical" means (topologically) generic – a property is *generic* if the set of elements possessing the property contains a residual subset, where residual is understood in the sense of Baire category: a residual subset is one that is a countable intersection of open, dense sets. Of course, this approach is only interesting if the underlying space has the *Baire property*, that a residual set is necessarily dense.

Other approaches to the question of describing "typical" dynamics are widespread, in particular those using measure theory. It is well known that the topological approach and the measure theoretic approach can give results that are superficially at odds – phenomena that are 'typical' in one sense are not 'typical' in the other. The tools we use are topological, and so give results in terms of residual subsets. The question of whether something analogous to our results can be obtained in the measure theoretic category is an open one.

The "typical" homeomorphism that we describe has properties that are quite different from the properties that are known to be "typical" for diffeomorphisms of smooth manifolds. The reason is that the usual topology on the set of diffeomorphisms includes information about derivatives, which of course is missing in the topology on the set of homeomorphisms. When dealing with homeomorphisms, there is very little fine-scale structure that cannot be destroyed — essentially anything other than general location can be perturbed away. On the other hand, with diffeomorphisms hyperbolic structures cannot be destroyed by perturbations in the smooth category. So, for instance, our "typical" homeomorphisms of the circle are very much different from the "typical" diffeomorphisms, which, as described in Nitecki (1971), Robinson (1995), or Katok and Hasselblatt (1995), have finitely many periodic orbits, with all other orbits asymptotic in both directions to one of the periodic ones.

Thus far we have put the word 'typical' in quotes to warn the reader that we have not been using it in its most common mathematical sense, namely a property that hold with probability 1. In what follows we will avoid using the term, but will speak of 'generic' properties and residual subsets. While the topological use of these terms that we employ is widespread, there are other uses of these terms (particularly in measure-theoretical studies) that are common. We regret any confusion that may result, but do not see any reasonable way to avoid it, other than to reiterate that all of our results are to be understood in a strictly topological sense.

0.2. Description of results

Let X denote a compact, metrizable space. The group of homeomorphisms on X, denoted $H(X)$, is a Polish group with respect to the topology of uniform convergence. That is, $H(X)$ is a topological group whose topology is separable and completely metrizable. In particular, and most important to us, is that $H(X)$ satisfies the *Baire property*: any countable intersection of open dense sets is necessarily dense; such an intersection (a dense G_δ) is called *residual*. We will say that a property P is satisfied by a *generic* homeomorphism on X if the set H_P of $f \in H(X)$ which satisfy P contains a residual set. For the most part, we are interested in generic properties that are conjugacy invariant, in other words, properties P such that $f \in H_P$ implies $hfh^{-1} \in H_P$ for all $h \in H(X)$. The latter condition says that the property P is preserved by continuous isomorphism between dynamical systems, where we regard a homeomorphism f on X as a discrete time dynamical system by iteration.

When X is a compact, piecewise linear manifold (and *a fortiori* if X is a compact smooth manifold) with $\partial X = \emptyset$ and dimension at least two, then, as we will show, a generic homeomorphism on X has a number of peculiar properties. Some of these properties carry over when the dimension is one, but some do not. For the present introductory discussion, assume that the dimension is two or more.

(1) We define the terms *attractor* and *repellor* in a strictly topological way, following Conley (1978); more details can be found in section 1. Suppose that f is generic and that A is an attractor for f.
   (a) A contains infinitely many repellors for f.
   (b) The interior of A is nonempty.
   (c) A contains repellors; in fact the interior of A is exactly the union of the basins of repulsion for the repellors contained in A.
   (d) The topological boundary of A, ∂A, is a quasi-attractor, that is, it is the intersection of a sequence of attractors, but is not itself an attractor (as ∂A has empty interior). The analogous statement is true for repellors.

   (e) It follows from the above that there are uncountably many distinct sequences $A_1 \supset R_1 \supset A_2 \supset R_2 ...$ with $A_1 = A$, the A_i's attractors and the R_i's repellors.
(2) Let $\mathcal{C}(f)$ denote the chain recurrent set of f, as defined in Conley (1978). For a generic homeomorphism f:
   (a) $\mathcal{C}(f)$ is a Cantor set.
   (b) In $\mathcal{C}(f)$, the set of periodic points is a dense subset of first category, i.e. the complement of the set of periodic points in $\mathcal{C}(f)$ is residual in $\mathcal{C}(f)$.
   (c) In particular, the chain recurrent set coincides with the nonwandering set.
(3) For a homeomorphism f a *chain component* B is by definition a maximal subset of $\mathcal{C}(f)$ such that any two points of B can be joined by ϵ-chains for any positive ϵ. The chain components are closed sets that partition the chain recurrent set. For generic f:
   (a) There are uncountably many chain components.
   (b) There is a dense set of points in $\mathcal{C}(f)$ which are contained in chain components on which f restricts to a system that has a nontrivial subshift of finite type as a factor. For brevity, we will refer to these points as "subshift type points".
   (c) A chain component B is called *terminal* if it has the property that for each $\epsilon > 0$, there is a $\delta > 0$ such that no δ-chain beginning in B can leave the ϵ-neighborhood of B. The restriction of f to a terminal chain component is either a single periodic orbit or is conjugate to a so-called "adding machine", a translation on a profinite group like the two-adic integers. In the latter case, the dynamics of f on B appear to be periodic when viewed to any finite resolution, although in fact B is not a periodic orbit (as the viewing resolution approaches 0, the apparent period goes to infinity). We will refer to the points of such a chain component as "adding machine points". The name comes from an analogy to a machine that has a sequence of registers, each of which can hold an integer from 0 up to some finite value, which can vary with the register. The process of adding 1 to the lowest register, 'carrying' to higher registers when a lower register 'turns over', defines a dynamical system, which looks periodic if only a finite number of registers are viewed. The situation is analogous to an odometer with an infinite number of dials, not all of which have the same length. In fact, what has recently been called an adding machine in the dynamics literature has been called an odometer in ergodic theory literature.
   For a dynamical example of such a set, imagine a finite collection of $n_1 > 1$ pairwise disjoint closed disks in the plane with each one mapped by f into the next and the last one mapped into the interior of the first, which we label $D_{1,1}$. Inside $D_{1,1}$ imagine there are $n_2 > 1$ pairwise disjoint closed disks, each mapped into the next by f^{n_1}, with the last being sent into the interior of the first, which is called $D_{1,2}$. Continue inductively in this way, and then consider the intersection

of the orbits of all of the $D_{1,j}$

$$A \equiv \bigcap_{j=1}^{\infty} \bigcup_{n=0}^{\infty} f^n(D_{1,j}) = \bigcap_{j=1}^{\infty} \bigcup_{n=0}^{n_j} f^n(D_{1,j}).$$

Then $f|A$ is an example of an adding machine chain component.

(d) The set of points of $\mathcal{C}(f)$ that lie in terminal chain components that are not periodic orbits form a residual subset of $\mathcal{C}(f)$. This set is disjoint from the subshift type points as well as from the periodic points. The same results hold for initial chain components, i.e. the terminal chain components for f^{-1}. In particular, the "dynamically isolated" points, which lie in chain components which are both initial and terminal, form a residual subset of the chain recurrent set.

(4) In general, the omega limit set of a point x is the set of limit points of the orbit sequence $\{f^n(x)\}$ as n tends to $+\infty$. Similarly, we obtain the alpha limit set of x by letting n tend to $-\infty$. For a generic homeomorphism f there is a residual subset of X consisting of points whose omega limit set is a terminal adding machine chain component and whose alpha limit set is an initial adding machine chain component. (B is an *initial* chain component for f if it is terminal for f^{-1}.) For such a point x, the two chain components are distinct, so x is not one of the 'dynamically isolated' points described in (3d).

The countable infinity of tiny attractors containing uncountably many chain components is an indication of great complexity in the dynamics of a generic homeomorphism. Suppose we look at such a system with the acuity of our vision such that we cannot distinguish between points which are ϵ-close for some positive ϵ. An adding machine subset then appears to be a periodic orbit. The residual set of points described in (4) then appear to be approaching periodic orbits in each direction. When we blow up the scale then what we previously regarded as an attracting orbit of period k is now seen to be a periodic set with interior. Within that interior most of the points are being attracted to and repelled from much smaller sets, mostly toward what seem to be periodic points of period some multiple of k. This picture, too, disintegrates when we increase the scale still further.

In general, it is hard to describe in further detail what these maps look like. However, the one dimensional case, where X is the circle, exhibits some of these complications in a fashion which can be easily described. Here we will look at the case where f is a generic homeomorphism of the circle. Such a homeomorphism has a periodic point, for this discussion we assume that it has a fixed point.

Let K be a Cantor set in $I = [0, 1]$ with $0, 1 \in K$. Let L be a smooth real-valued map on I vanishing exactly on K and let f be the time one map of the flow for $dx/dt = L(x)$. So $I \backslash K$ consists of infinitely many disjoint, invariant, open intervals with f moving all points on each one of these intervals of in the same direction, depending upon whether L was positive or negative on the interval. Call these "up intervals" and "down intervals", respectively. Assume that between any two of the intervals there are down intervals as well as up intervals. Now identify the points 0 and 1 to get a homeomorphism of the circle. If we remove any up interval and any down interval the circle breaks into two closed intervals. On the circle, one of these intervals, A, has the property that any orbit in a neighborhood of it will approach A, while the other interval, R, has the property that any backward orbit in a neighborhood of it will approach R. This pair (A, R) is an example of what is called an *attractor-repellor pair* for f, in that any point not in $A \cup R$ approaches A under iteration by f, and it approaches R under iteration by f^{-1}. A little contemplation verifies the conditions in (1) and (2). The chain recurrent set is the set of fixed points K and the chain components are the individual points of K. Except for the points that are endpoints of the open intervals complementary to the Cantor set, each chain recurrent point is both initial and terminal. Every non fixed point has its omega limit set a fixed point that is terminal but not initial, and its alpha limit set is a fixed point that is initial but not terminal; the union of all these limit sets is the collection of endpoints of the arcs complementary to the Cantor set $\mathcal{C}(f)$. We will later see that among orientation preserving maps of the circle with a fixed point this is the generic picture. However, because the dimension is one, we don't see the complex behavior on the individual chain components described by the conditions in (3) above.

Now we return to the general discussion of manifolds with dimension bigger than 1. In (2) and (3) we described three different types of behavior, all occurring on dense subsets of the chain recurrent set: periodic motion, the somewhat regular but aperiodic motion of the "adding machine points", and the irregular motion of the subshift type points. Somewhat surprisingly, for generic f, a residual subset of X consists of points that are Lyapunov stable. In fact, each of these points x satisfies an even stronger stability condition called *chain continuity* (Akin 1996), meaning that for each $\epsilon > 0$ there is a $\delta > 0$ with the property that if (x_n) is a δ-chain for f with $x_0 = x$, then the distance from x_n to $f^n(x)$ is less than ϵ for every $n > 0$. It turns out that a point is a chain continuity point exactly when its omega limit set is a terminal chain component on which the map is either a periodic orbit or an adding machine. Thus, from (3d) and (4) follows:

(5) For a generic homeomorphism f the points which are chain continuous for f and for f^{-1} form a residual subset of X which intersects $\mathcal{C}(f)$ in a residual subset of $\mathcal{C}(f)$.

In general, a homeomorphism f in $H(X)$ is called *almost equicontinuous* when the set of points which are equicontinuity points for f and for f^{-1} is residual in X. So condition (5) says that a generic homeomorphism is almost equicontinuous.

Thus, the dynamics of a generic homeomorphism are geometrically complicated but the behavior of most (in the topological sense) of the individual orbits is quite stable.

0.3. Brief remarks about techniques

As we have mentioned, our techniques and results are topological. The same questions that we address can also be asked in terms of measure theory. While there are difficulties in finding an appropriate measure on the space of homeomorphisms, at least on a manifold Lebesgue measure is certainly appropriate in the context of questions of the behavior of "most" points. As is well known, it is likely that results in the two settings will be somewhat different, perhaps even evidently contradictory – recall the fact that a compact manifold can be written as the union of a set of first Baire category and a set of Lebesgue measure 0 (see Oxtoby (1980) for a nice discussion of these matters). We remind the reader of another two examples showing that metrically-typical may be quite different from topologically-typical in dynamical systems. Herman (1977) showed that if f_t, $0 \leq t \leq 1$ is a C^1 path of orientation preserving C^r diffeomorphisms of the circle ($r \geq 3$), such that the rotation number is not constant along the path, then for each t in a set of positive Lebesgue measure, f_t is C^{r-2} conjugate to an irrational rotation. Jakobson (1980) showed that there is a set with positive Lebesgue measure consisting of parameters μ in the interval $[4-\epsilon, 4]$ such that the map $x \mapsto \mu x(1-x)$ has an invariant measure absolutely continuous with respect to Lebesgue measure.

Because we are working in the continuous category rather than the smooth, the types of perturbations we can employ are fairly violent. There are three general types of perturbation that we employ: 'closing', where a periodic orbit is created out of an approximately periodic orbit; 'trapping', where a set is mapped well into its interior; and 'period multiplication', used to create periodic phenomena of arbitrarily high period. At the level of individual periodic orbits, these ideas are fairly straightforward. On the other hand, individual periodic orbits are easy to destroy using C^0 perturbations. In order to describe open dense sets of homeomorphisms it is necessary to find properties that are "immune" to such perturbations. This is perhaps the main difficulty that was encountered in early attempts to show that C^0 generically, periodic points are dense in the nonwandering set; see Hurley (1996b) for a brief description and further references. Thus, in our constructions we do not focus on individual orbits; instead we look at how certain sets are mapped, with two goals in mind: we want the way that they are mapped to be stable under small perturbation – a closed set that is mapped into its interior is the basic example, and we want them to be mapped in such a way that we can extract some information about orbits of individual points from the gross information about the map on the set – here the basic example is a Euclidean disk being mapped into itself, implying the existence of a fixed point. Working with sets, with these two goals, means that some of our definitions and notations become a bit cumbersome.

Additionally, we have tried to frame our results so that they apply to fairly general compact metric spaces, not just manifolds. Because of this, we spend some time developing the hypotheses that allow us to make the type of perturbations we need. For the most part, these technical hypotheses are easily seen to hold in the case of a manifold.

0.4. Comparison with the smooth case

On a smooth manifold X our C^0-generic homeomorphisms are very different from C^1-generic diffeomorphisms, This is not surprising – for one thing, lack of differentiability is itself C^0-generic. A basic C^1 result is the Kupka-Smale theorem, which says in part that a C^1-generic diffeomorphism has all of its periodic points hyperbolic, with the consequence that it has only finitely many periodic points of any given period. On the other hand, the topologically generic homeomorphism has infinitely many points of some period (Palis *et. al.*, 1975, Hurley, 1996b).

Let us compare our picture of a generic homeomorphism on X with some standard smooth examples, namely Morse-Smale diffeomorphisms and Anosov diffeomorphisms (descriptions of these diffeomorphisms can be found in many places, for example Robinson (1995) and Katok and Hasselblatt (1995)).

0.4.1. Perturbations of Morse-Smale diffeomorphisms. First let us look at a local perturbation in one dimension. This will provide at least partial illumination. Perturb a parabola with a single maximum to a C^∞ function which is C^0 close but which has a tiny Cantor set of critical points. The gradient flow of the perturbed function homes in toward this little Cantor set. Of course, in this one dimensional case the behavior near the Cantor set is simple as well. It looks like the fixed point case for circle homeomorphisms.

For the remainder of this subsection, assume that X is a smooth manifold without boundary of dimension $n \geq 2$.

Consider a Morse-Smale diffeomorphism, or more generally, let g be any homeomorphism on X whose chain recurrent set $\mathcal{C}(g)$ is finite. Take an open set O, the union of a collection of pairwise disjoint open topological balls, each meeting $\mathcal{C}(g)$ in a single point, so that $\mathcal{C}(g) \subset O$. The map $h \mapsto \mathcal{C}(h)$ is upper semicontinuous, which implies that $\mathcal{C}(f) \subset O$ as long as f is close enough to g. If f is one of our generic homeomorphisms with $d(f,g)$ so small that $\mathcal{C}(f) \subset O$, then all of the complicated structure of the chain recurrent set of f is contained in the disjoint union of the finitely many small balls, O.

0.4.2. Perturbations of Anosov diffeomorphisms. Because in the Morse-Smale case the chain recurrent set started out small and stayed locally small in diameter under perturbation, the g picture provides reasonable intuition for the f behavior. For Anosov diffeomorphisms the situation is completely different.

It is well known that an Anosov diffeomorphism g on X is *structurally stable*. That is, if f is a diffeomorphism on X which is close enough to g in the C^1 sense then f is an Anosov diffeomorphism and f is conjugate to g by some homeomorphism h. The Anosov diffeomorphism g is also *topologically stable*: if f is a homeomorphism on X which is close enough to g in the C^0, uniform, metric then there is a surjective continuous map h on X mapping f to g, i.e. there is a semi-conjugacy h from f to g. In particular, there is such an f satisfying our genericity conditions (1) - (5).

In many ways, a transitive Anosov diffeomorphism g is exactly the opposite of a generic homeomorphism f. Since g is transitive, the entire manifold X is a single chain component for g. For f the chain recurrent set is a Cantor set in X consisting of uncountably many distinct chain components. On the other hand, f is almost equicontinuous while g is expansive. It is hard to imagine the semiconjugacy mapping f to g. In fact, h is a profoundly singular map. We will show (Theorem 11.8) that there is an open, dense, f-invariant subset of X on which h is locally constant, and there is a single chain component B for f with h mapping B onto the entire manifold X.

0.5. Standing notation

Throughout this *Memoir* our spaces X, Y etc are all assumed to be compact metric spaces unless otherwise mentioned. The major exceptions – explicitly indicated – will be Polish spaces, i.e. separable, completely metrizable spaces, like $H(X)$ and its closed subgroups. We will use $d(\cdot,\cdot)$ to denote the metric. We will denote the closure, interior, and topological boundary of a subset $B \subset X$ by $\overline{B}$, B°, and $\partial B \equiv \overline{B} \backslash B^\circ$, respectively. The set of all points that are within distance ϵ of B is denoted $V_\epsilon(B)$

$$V_\epsilon(B) \underset{\text{def}}{=} \{x \mid d(x,B) < \epsilon\ \}.$$

A Cantor set is a perfect, zero-dimensional, compact metric space. A manifold X is a compact topological manifold (and so of finite dimension) with boundary ∂X. The prefix p.l. means piecewise linear and smooth means C^∞. Recall that any smooth manifold can be smoothly triangulated to get a p.l. manifold, see e.g. Munkres (1963).

Our approach in this work is almost exclusively topological, and this is reflected in our choice of terminology. Definitions of terms, such as 'attractor' and 'repellor', are purely topological; the definitions are not original with us, and appear commonly in the literature (see Conley (1978), Pilyugin (1994), Katok and Hasselblatt (1995)), but one should be aware that other definitions are also common, some requiring additional topological structure (Robinson (1995)), others requiring less (Bhatia and Szegö (1967)), and others requiring measure-theoretic considerations (Guckenheimer and Holmes (1983), Milnor(1985)).

0.6. Road map

In this introduction we have concentrated on describing how our results apply to homeomorphisms of manifolds. Our general results apply to a much wider collection of compact spaces than manifolds. We require some technical assumptions, so as to be able to make certain perturbations. The reader who is only interested in the manifold case may find the discussion of the technical details of the general case somewhat over-meticulous, but we thought it important to present the results in a general way. The technical assumptions that we require are all easily seen to hold if X is a manifold.

We now give a brief overview of the structure of the paper. The first 3 sections generally give background, develop notation, and introduce techniques that will be employed later in the paper. Sections 4 and 5 contain the basic perturbation results that we will exploit for most of our results. The main genericity theorems are in section 6, while section 7 makes the connection with almost equicontinuity. Sections 8 and 9 develop our general theory in the cases of a Cantor set, and the circle, respectively. In the last 3 sections we specialize to the case where X is a manifold; section 10 shows that generically the chain recurrent set is a Cantor set, section 11 outlines our results for homeomorphisms on a manifold, and section 12 describes what our techniques show for continuous maps on a manifold that are not invertible.

0.7. Comments

Recent results of Pilyugin and Plamenevskaya (1999) complement our work. They prove that the *Shadowing Property*, also called the *Pseudo-Orbit Tracing Property*, is generic. See also Pilyugin (1999).

One also sees theorems about the "generic homeomorphism" on a manifold in the context of the space of homeomorphisms which preserve a given Lebesgue type measure on a connected manifold. Here the Oxtoby-Ulam Theorem from Oxtoby and Ulam (1941) says that the generic homeomorphism is ergodic. In particular, it is topologically transitive and so the whole space is a single chain component. For a beautiful recent exposition of these contrasting results, see Alpern and Prasad (2000).

This paper extends earlier work by some of us [Kennedy (1996); Hurley (1995), (1996a)] to describe the dynamic behavior of a generic homeomorphism on a compact, piecewise linear manifold.

We thank our colleagues and friends for helpful discussions about these matters. We are also grateful to Kate March for her preparation of the original manuscript.

CHAPTER 1

Attractors and Chain Recurrence

Overview: The purpose of this section is to review some basic results about attractors of maps on compact metric spaces, and their connection with the concept of chain recurrence. We describe a natural equivalence relation on the chain recurrent set, the equivalence classes are called chain components. There are several ways of topologizing the space of chain components, which we describe.

Our work builds upon the topological theory of attractors and its connections with the idea of chain recurrence, as developed by Conley (1978). We will use Akin (1993), hereafter referred to as GTDS, as a general reference for the many folklore results we will need. Other general references are Nitecki (1971), Robinson (1995), and Katok and Hasselblatt (1995). We begin by reviewing some basic definitions and introducing our notation. Recall that our topological spaces are assumed to be compact, metric spaces.

If $f : X \to X$ is a continuous map and $x \in X$, then the forward iterates f^n, $n \geq 0$ of f are defined by $f^0(x) = x$ and $f^n(x) = f(f^{n-1}(x))$ for each $n \geq 1$. The *forward orbit* of $p \in X$ is the sequence $f^n(p)$, and is denoted by $\mathcal{O}(x, f)$. If $A \subset X$ is nonempty, then its *omega-limit set* under f is

$$\omega(A, f) \underset{\text{def}}{=} \limsup f^n(A) \underset{\text{def}}{=} \bigcap_k \overline{\bigcup_{n\geq k} f^n(A)}. \tag{1.1}$$

Observe that if U is any neighborhood of $\omega(A, f)$ then $f^n(A) \subset U$ for all sufficiently large n. The omega limit set is the smallest closed set with this property. If A is just a single point $\{p\}$, then we denote its omega limit set simply as $\omega(p, f)$. Note that $\overline{\mathcal{O}(A, f)} = \mathcal{O}(A, f) \cup \omega(A, f)$. When f is a homeomorphism, we also define its *alpha limit set* by

$$\alpha(A, f) = \omega(A, f^{-1}).$$

A subset A is called *forward invariant* under a map g on X (or just g-forward invariant) if $g(A) \subset A$. A is *g-invariant* if $g(A) = A$. A subset B is called *g^{-1}-invariant* if $B = g^{-1}(B)$. Notice that

$$A = g^{-1}(A) \Rightarrow g(A) = gg^{-1}(A) \subset A. \tag{1.2}$$

That is, a g^{-1}-invariant set is g-forward invariant. Furthermore, it is g-invariant when the mapping g is surjective because then $gg^{-1}(A) = A$.

Note that $\omega(A, f)$ is the largest invariant set contained in $\overline{\mathcal{O}(A, f)}$.

1.1. Chain recurrence

A major part of our investigation concerns Conley's *chain recurrent set*, whose definition follows. Given $\epsilon > 0$, an *ϵ-chain* for f is a finite sequence $x_0, x_1, \cdots, x_n$, with $n > 0$ and $d(f(x_{j-1}), x_j) < \epsilon$ for all $0 < j < n$ (d is the metric on X). This

ϵ-chain is said to *begin* at x_0 and *end* at x_n, and the integer n is its *length*. We denote the set of points that are ends of ϵ-chains beginning at x by $\mathcal{C}_\epsilon(x, f)$, and define

$$\mathcal{C}(x, f) \underset{\text{def}}{=} \bigcap_{\epsilon>0} \mathcal{C}_\epsilon(x, f). \tag{1.3}$$

If $0 < \delta < \epsilon$, then it is easy to verify that $\overline{\mathcal{C}_\delta(x, f)} \subset \mathcal{C}_\epsilon(x, f)$, from which it follows that $\mathcal{C}(x, f)$ is closed. If $B \subset X$, then $\mathcal{C}_\epsilon(B, f)$ (resp. $\mathcal{C}(B, f)$) denotes the union of $\mathcal{C}_\epsilon(x, f)$ (resp. $\mathcal{C}(x, f)$) for all $x \in B$. A point $p \in X$ is called *chain recurrent* if $p \in \mathcal{C}(p, f)$. The set of all chain recurrent points of f is denoted by $\mathcal{C}(f)$. Standard facts about the chain recurrent set are that it is compact, nonempty, and that if f is a homeomorphism, then $\mathcal{C}(f^{-1}) = \mathcal{C}(f)$. Clearly all periodic points are chain recurrent. The chain recurrent set also contains all points that are *recurrent* (i.e., contained in their own omega limit sets) or *nonwandering* (a point p is nonwandering if each neighborhood of p contains a point x such that the forward orbit of $f(x)$ intersects the neighborhood – note that this is essentially the same as the definition of chain recurrence, but allowing jumps from $f(x_{j-1})$ to x_j only at the beginning and the end of ϵ-chains). The nonwandering set is denoted by $\Omega(f)$.

In short, $\mathcal{C}(x, f)$ is the set of points y such that for each $\epsilon > 0$ there is an ϵ-chain starting at x and ending at y. Note that there is an obvious transitivity. By concatenating ϵ-chains, we see that if $y \in \mathcal{C}(x, f)$ and $z \in \mathcal{C}(y, f)$, then $z \in \mathcal{C}(x, f)$. It is a simple consequence of continuity that if p is a chain recurrent point, then $p \in \mathcal{C}(f(p), f)$. In particular, since it is obvious that $f(p) \in \mathcal{C}(p, f)$, we see that

$$p \in \mathcal{C}(f) \Leftrightarrow p \in \mathcal{C}(f(p), f) \Leftrightarrow f(\mathcal{C}(p, f)) = \mathcal{C}(p, f). \tag{1.4}$$

One defines an equivalence relation $\sim$ on $\mathcal{C}(f)$ by

$$x \sim y \text{ if } x \in \mathcal{C}(y, f) \text{ and } y \in \mathcal{C}(x, f). \tag{1.5}$$

The associated equivalence classes are called the *chain components* for f. It is easy to see that each chain component is closed, and that when f is a homeomorphism, the chain components of f^{-1} are the same as those of f.

We will also speak of $\mathcal{C}(f)$*-invariant sets*; such a set is defined by the property that it contains $\mathcal{C}(x, f)$ for every x in the set.

It will useful to define an analog of omega limit sets for ϵ-chains. We define the *chain omega limit set* of a set A to be set of points $x \in X$ with the property that for any $\epsilon > 0$ and any integer n, there is an ϵ-chain of length at least n that begins in A and ends at x. This set is denoted $\omega\mathcal{C}(A, f)$, or simply as $\omega\mathcal{C}(x, f)$ if $A = \{x\}$. Note that

$$p \in \mathcal{C}(f) \cap \mathcal{C}(x, f) \Rightarrow p \in \omega\mathcal{C}(x, f), \tag{1.6}$$

because we can follow an ϵ-chain from x to p by one of arbitrary length from p to p. It is a simple consequence of continuity that $\mathcal{C}(x, f) \backslash \omega\mathcal{C}(x, f) \subset \mathcal{O}(x, f)$; i.e., that

$$\mathcal{C}(x, f) = \mathcal{O}(x, f) \cup \omega\mathcal{C}(x, f).$$

It follows that

$$\omega\mathcal{C}(x, f) \text{ is the largest invariant subset of } \mathcal{C}(x, f), \tag{1.7}$$

to see this, note that $\omega\mathcal{C}(x, f)$ is clearly forward invariant, and the previous equation shows that there is no larger invariant subset of $\mathcal{C}(x, f)$, so we only need to verify that it is invariant. To do this, suppose $y \in \omega\mathcal{C}(x, f)$. For each $n \geq 1$, let γ_n be a finite $(1/n)$-chain of length at least n that begins at x and ends at y, and let $z(n)$ be the next to last point in γ_n. If z is an accumulation point of the sequence $z(n)$, then it is easy to verify that $f(z) = y$ and that $z \in \omega\mathcal{C}(x, f)$.

A continuous map f is called *chain transitive* when the entire space is a single chain component. A subset A is called a *chain transitive subset* if A is a nonempty, closed, invariant subset of X such that the restricted map $f_A : A \to A$ is chain transitive. Each chain component and each limit point set $\omega(x, f)$ is a chain transitive subset, see GTDS, Theorem 4.12 and Proposition 4.14. On the other hand it is clear that any chain transitive subset is contained in a chain component.

1.2. Attractors

There is an important connection between the structure of the set of chain components of f and the set of attractors of f. In this, "attractor" is defined topologically, as in in Conley (1978). To describe this, we first define the *strong inclusion* symbol:

$$A_1 \subset\subset A_2 \Leftrightarrow \overline{A_1} \subset A_2^\circ. \tag{1.8}$$

A closed set U is called *inward* for f when

$$f(U) \subset\subset U. \tag{1.9}$$

A set U which satisfies (1.9) is sometimes called a *trapping region*, so the closure of any trapping region is inward. If the interior of U contains the ϵ-neighborhood of $f(U)$ then any ϵ-chain which begins in U terminates in U°. So if U is inward for f then

$$\mathcal{C}(U, f) \subset\subset U^\circ. \tag{1.10}$$

If U is an inward set for f then the associated *attractor* A for f, is defined to be $\omega(U, f)$, which in this case is simply $\cap_n f^n(U)$.

Thus an attractor A for f is a closed f-invariant set which is the limit of the decreasing sequence of iterates $\{f^n(U)\}$ for some inward set U. Note that for such a U and A, for any set K with $A \subset K \subset U$, then

$$\omega(K, f) = A. \tag{1.11}$$

Because the compact metric space X has a countable basis, and a finite number of basic open sets cover any attractor, it follows that the number of distinct attractors is no more than the number of finite unions of the basic open sets, so the number of attractors is at most countable.

We will also speak of sets as being "inward for f^{-1}"; of course if f is a homeomorphism, this is defined exactly as above. In the case that f is not invertible, saying that a set U is *inward for* f^{-1} means that

$$f^{-1}(U) \subset\subset U,$$

understood in the sense of inverse images. It is easy to check that if U is inward for f, then the complementary set $X\backslash U$ is inward for f^{-1}.

If U is inward for a homeomorphism f, then $f(U)$ is also inward for f. When f is not invertible, this may no longer be the case, however, by slightly 'fattening' the image of U we can get another inward set:

(1.12) If U is inward for f, then so is $V_\epsilon(f(U))$, provided that $\epsilon > 0$ is small enough that $V_\epsilon(f(U)) \subset U^\circ$.

A central observation, due originally to Conley, is that if U is inward for f with attractor A, then any chain recurrent point that is contained in U must actually lie in A. The reason is simple; we have remarked already that no point of $U \backslash f(U)$ can be chain recurrent for f. The same argument shows that if $n \geq 1$ then no point of $U \backslash f^n(U)$ is chain recurrent for f^n, so it is only necessary to prove that f and f^n have the same chain recurrent sets.

PROPOSITION 1.1. *Let f be a continuous map on X and k be a positive integer.*

$$\mathcal{C}(f) = \mathcal{C}(f^k). \tag{1.13}$$

Moreover, if B_0 is a chain component of f^k, then $B \underset{\text{def}}{=} \bigcup_{i=0}^{k-1} f^i(B_0)$ is the chain component of f containing B_0.

PROOF. It is obvious that one can "fill out" an ϵ-chain for f^k to obtain one for f with the same beginning and end (just interpolate the jth iterates of points in the chain for f^k, $1 \leq j < k$). Hence $\mathcal{C}(f^k) \subset \mathcal{C}(f)$. To obtain the opposite inclusion, note that by uniform continuity of f, for a given $\epsilon > 0$ there is a $\delta > 0$ such that if (x_j) is a δ-chain for f of length nk, then $(x_{ik})_{i=0}^n$ is an ϵ-chain for f^k of length n. This establishes (1.13).

The argument above also shows that B_0 lies in a single chain component of f; call this chain component B'. Chain components are invariant, so B' contains B. Let $b \in B_0$, and suppose that $z \in B'$. For each $n > 0$ there is a $(1/n)$-chain for f, γ_n, of length L_n, that begins and ends at b and contains z. By concatenating such a chain with itself an appropriate number of times, we obtain a chain whose length is a multiple of k, so we may as well assume that each L_n is a multiple of k. Let M_n denote the length of some initial segment of γ_n that ends at z; the concatenation argument allows us to assume that L_n is much larger than M_n, which in turn is much larger than k. For some j with $1 \leq j \leq k$ there are infinitely many n's for which M_n is equivalent to j modulo k. Let y_n denote the point in γ_n that occurs in position $M_n - j$; we can assume that y_n converges to some point y. By continuity, $f^j(y) = z$, and by the arguments of the last paragraph, we see that y is in the same chain component of f^k as b, namely B_0. But this implies that $z \in f^j(B_0) \subset B$, which finishes the proof that $B = B'$. □

COROLLARY 1.2. *If U is inward for f and $x \in U \cap \mathcal{C}(f)$, then x lies in the attractor determined by U, namely $\omega(U, f)$. In particular, if A is an attractor, then $\mathcal{C}(x, f) \subset A$ for each $x \in A$.*

PROOF. The first sentence follows from the Proposition as in the discussion leading up to the Proposition. The second follows from the combination of the first and (1.12), because (1.12) and induction shows that any neighborhood of A contains an inward set whose omega limit set is A. □

An important collection of examples of inward sets are the sets $\overline{\mathcal{C}_\epsilon(x, f)}$, where $x \in X$ and $\epsilon > 0$ are arbitrary (see (1.3)). To see that these sets are inward, note that it follows directly from the definitions that if $y \in \mathcal{C}_\epsilon(x, f)$ then $V_\epsilon(f(y)) \subset \mathcal{C}_\epsilon(x, f)$, showing that $\mathcal{C}_\epsilon(x, f)$ satisfies (1.9). Thus

$$\text{For any } x \in X \text{ and } \epsilon > 0,\ \overline{\mathcal{C}_\epsilon(x, f)} \text{ is inward.} \tag{1.14}$$

If we let $A_\epsilon(x)$ denote the corresponding attractor $\omega(\overline{\mathcal{C}_\epsilon(x,f)}, f)$ then $A_\epsilon(x)$ is the largest invariant set in $\overline{\mathcal{C}_\epsilon(x,f)}$, so it contains $\omega\mathcal{C}(x,f)$; moreover,

$$\bigcap_\epsilon A_\epsilon(x) \subset \bigcap_\epsilon \overline{\mathcal{C}_\epsilon(x,f)} = \mathcal{C}(x,f)$$

and $\cap A_\epsilon(x)$ is invariant, so $\cap_\epsilon A_\epsilon(x) \subset \omega\mathcal{C}(x,f)$ by (1.7). The two inclusions show that

$$\bigcap_\epsilon A_\epsilon(x) = \omega\mathcal{C}(x,f). \tag{1.15}$$

1.3. Attractor-repellor pairs

Suppose U is an inward set for f with associated attractor A. Let W denote the union of all inverse images of U under f

$$W = \bigcup_{n\geq 0} f^{-n}(U),$$

and let R be the complement of W, $R \underset{\text{def}}{=} X\backslash W$, so

$$R = \{x \in X \mid \mathcal{O}(x) \cap U = \emptyset\}. \tag{1.16}$$

The three sets, A, R, and $W\backslash A$ form a piecewise disjoint cover of X. R is called the *repellor dual* to the attractor A, and the pair (A, R) is called an *attractor-repellor* pair. The open set $W = X\backslash R$ is called the *basin of attraction* for A, with $W\backslash A = X\backslash(A \cup R)$ the *proper basin* for the pair. Note that if $p \in W\backslash A$ then p is not chain recurrent: Corollary 1.2 shows that $U\backslash A$ is disjoint from $\mathcal{C}(f)$; if p is in $W\backslash U$ then there is a positive n with $f^n(p) \in U^\circ$. For this n and sufficiently small $\epsilon > 0$, we have both (i) any length n ϵ-chain starting at p ends in U°, and (ii) any ϵ-chain starting in U ends in U. It follows that no ϵ-chain of length at least n can both start and end at p, so p is not chain recurrent.

Thus the attractor-repellor pair (A, R) is characterized as a pair of closed invariant sets which satisfy the conditions:

$$(i)\ A \cap R = \emptyset, \quad (ii)\ \mathcal{C}(f) \subset A \cup R, \quad (iii)\ \mathcal{C}(f|A) = A, \quad (iv)\ \mathcal{C}(f^{-1}|R) = R. \tag{1.17}$$

It follows that if B is a chain component that meets $X\backslash R$, then $B \subset A$. Similarly, if B is a chain component that meets the complement of A, then B is contained in the dual repellor R.

If A, R is an attractor-repellor pair for f then the attractor A is f-invariant. Because the repellor R is f^{-1}-invariant, (1.2) implies that R is f forward invariant and is f-invariant when f is surjective. For example, when f is a homeomorphism, f-invariance is equivalent to f^{-1}-invariance.

Clearly, if K is a compact subset of the basin of attraction of A then

$$\omega\mathcal{C}(K, f) \subset A. \tag{1.18}$$

We saw in (1.15) that any chain omega limit set is an intersection of attractors. A nonempty invariant set that is the intersection of finitely many attractors is easily seen to be an attractor, but a nonempty invariant set that is the intersection of infinitely many distinct attractors is not an attractor. We call any nonempty invariant set that is an intersection of attractors a *quasi-attractor* (note that an attractor is also a quasi-attractor). Thus, by (1.15)

(1.19) Any chain omega limit set is a quasi-attractor.

There is another way of characterizing quasi-attractors. Corollary 1.2 shows that if A is an attractor, then $\mathcal{C}(x, f) \subset A$ for all $x \in A$. Since a quasi-attractor is a nonempty intersection of attractors, it has the same property. Conversely, if B is a compact invariant set with this property, then B is a quasi-attractor. To see this, note that if W is any open neighborhood of B, then there is an $\epsilon > 0$ such that $\mathcal{C}_\epsilon(x, f) \subset W$ for all $x \in B$; if this were not so, then there would be points $x_n \in B$, $y_n \in X \backslash W$ with a $(1/n)$-chain from x_n to y_n for each n. By compactness we may assume that x_n converges to $x \in B$ and y_n converges to $y \in X \backslash W$, giving the contradiction that $y \in \mathcal{C}(x, f) \backslash B$. Since B is compact, for each ϵ it is covered by a finite number of the inward sets $\mathcal{C}_\epsilon(x, f)$; the union of a finite number of inward sets is inward, and the associated attractor A contains B. Thus for any neighborhood W of B there is an attractor A with $B \subset A \subset W$, from which it follows that B is a quasi-attractor. In short, for a compact invariant set Q,

$$Q \text{ is a quasi-attractor } \Leftrightarrow \mathcal{C}(x, f) \subset Q \quad \text{for all } x \in Q. \tag{1.20}$$

Note that in establishing (1.20) we have also shown that if Q is a quasi-attractor, then

$$\text{there is a neighborhood basis of } Q \text{ made up of open inward sets.} \tag{1.21}$$

1.4. Initial and terminal chain components

Recall the definition of a chain component, (1.5). The quotient map from $\mathcal{C}(f)$ to $\mathcal{B}_f \underset{\text{def}}{=} \mathcal{C}(f)/\sim$, the space of chain components, induces upon the latter the structure of a compact metrizable space.

There is a natural partial ordering on $\mathcal{B}_f$ defined by $B_1 \rightsquigarrow B_2$ if and only if $y \in \mathcal{C}(x, f)$ for $x \in B_1$ and $y \in B_2$ (the condition is clearly independent of the choice of the points x and y). We call $B \in \mathcal{B}_F$ *terminal* (resp. *initial*) if it is minimal (resp. maximal) with respect to this partial order. In other words, B is terminal if $B \rightsquigarrow B'$ implies that $B' = B$.

REMARK 1.1. We use the expression "terminal chain component" to avoid confusion with the usual concept of a *minimal set* for f, i.e. a minimal element in the family of nonempty, closed, f invariant subsets of X. A point of a minimal set will be called a *minimal point* The mapping f is called a *minimal map.* when X is a minimal set, i.e., when there is no proper, closed, invariant subset.

Note that there are several ways of characterizing when a chain component B containing a point x is terminal:

$$B \text{ is terminal } \Leftrightarrow \omega\mathcal{C}(x, f) = B \Leftrightarrow \omega\mathcal{C}(x, f) \text{ is chain transitive.} \tag{1.22}$$

In particular, it now follows from (1.19) that a terminal chain component is a quasi-attractor. We pause to collect some of these results in the following.

PROPOSITION 1.3. *Let f be a continuous map on X and B be a closed subset of X. The following conditions are equivalent.*

(1) *B is a terminal chain component.*

(2) *B is a chain transitive quasi-attractor.*

(3) *B is a minimal element, with respect to set inclusion, in the family of closed, nonempty, $\mathcal{C}f$-forward invariant subsets of X*

(i.e., those subsets K with the property that $\mathcal{C}(x,f) \subset K$ for each $x \in K$).

PROOF. (1) $\Rightarrow$ (2): The remarks immediately preceding the Propositions show that a terminal chain component is a quasi-attractor. As remarked above, any chain component is chain transitive.

(2) $\Rightarrow$ (3): Since B is a quasi-attractor it is in the family. If A is a family member contained in B with $x \in A$ then by $\mathcal{C}(f)$ forward invariance of A and chain transitivity of B:

$$B \subset \mathcal{C}(x,f) \subset \mathcal{C}(A,f) \subset A, \tag{1.23}$$

proving minimality.

(3) $\Rightarrow$ (1): If $x \in B$ then $\omega(x,f) \subset \mathcal{C}(x,f) \subset B$ because B is in the family. Since $\omega(x,f) \neq \emptyset$, $\mathcal{C}(x,f)$ is in the family as well. By minimality $\mathcal{C}(x,f) = B$ and so B is a terminal chain component. □

COROLLARY 1.4. *Every nonempty quasi-attractor for a continuous map f contains a terminal chain component for f.*

PROOF. Use the usual Zorn's Lemma argument to get a minimal nonempty, closed, $\mathcal{C}(f)$-forward invariant subset. □

A chain component is a quasi-attractor exactly when it is terminal. So a chain component B is an attractor iff it is terminal and $B \cap \mathcal{C}(f)$ is clopen in $\mathcal{C}(f)$ i.e. B is an isolated point in $\mathcal{B}_f$.

1.5. The space(s) of chain components

As in the last subsection, $\mathcal{B}_f$ denotes the space of chain components of f. A useful metric on $\mathcal{B}_f$ is obtained by using a complete Lyapunov function, which we now describe. A *Lyapunov function* for f is a continuous map $L : X \to \mathbf{R}$ such that $y \in \mathcal{C}(x,f)$ implies $L(y) \geq L(x)$. Such a function is constant on each chain component. A *complete Lyapunov function* distinguishes chain components and so induces a homeomorphism of $\mathcal{B}_f$ onto a compact subset of $\mathbf{R}$. It is always the case that complete Lyapunov functions $L : X \to \mathbf{R}$ exist with $L(\mathcal{C}(f))$ contained in a Cantor set. This result is sometimes called the *Fundamental Theorem of Dynamical Systems* (see GTDS Theorem 3.12). It follows that $\mathcal{B}_f$ is always zero-dimensional. We can use such an L to define a distance between chain components $B_1, B_2 \in \mathcal{B}_f$, namely $|L(B_1) - L(B_2)|$. We will allow context to determine whether we think of a chain component B as a subset of $\mathcal{C}(f)$ or as a point of $\mathcal{B}_f$. Note that $B_1 \rightsquigarrow B_2$ if and only if $L(B_1) \geq L(B_2)$, with equality if and only if $B_1 = B_2$.

Let $C(X)$ denote the set of closed subsets of X and define

$$\begin{cases} CT(f) = \{A \in C(X) : A \neq \emptyset \text{ is chain transitive}\} \\ \quad = \{A \in C(X)\backslash\emptyset : f(A) = A \text{ is a chain component of } f|A\}. \end{cases} \tag{1.24}$$

Recall that $C(X)$ is itself a compact metric space when equipped with the Hausdorff metric which is defined as follows:

$$\begin{cases} \emptyset \text{ is an isolated point of } C(X), \text{ and for nonempty } A_1, A_2 \in C(X), \\ d(A_1, A_2) = \inf\{\epsilon : A_1 \subset V_\epsilon(A_2) \text{ and } A_2 \subset V_\epsilon(A_1)\}, \end{cases} \tag{1.25}$$

where $V_\epsilon(A)$ is the set of points of X that are within ϵ of some point of A.

It is easy to check that $CT(f)$ is a closed subset of $C(X)$ (see GTDS Exercise 7.37 (d)). If f is a homeomorphism then A is chain transitive with respect to f^{-1} if it is so with respect to f, i.e. $CT(f) = CT(f^{-1})$ when f is a homeomorphism.

The identity map 1_X is chain transitive iff X is connected, see GTDS Exercise 1.9, and so $CT(1_X)$ is the space of closed, connected subsets of X. The chain components of 1_X are the components of X.

There is a surjection

$$\beta : CT(f) \longrightarrow \mathcal{B}_f \tag{1.26}$$

which associates to $A \in CT(f)$ the chain component $\beta(A)$ containing A. This map is continuous and maps each chain component B, regarded as an element of $CT(f)$, to B regarded as an element of $\mathcal{B}_f$. However, the topology on $\mathcal{B}_f$ is usually strictly coarser than that induced on the chain components from $CT(f)$ (see GTDS, Exercise 7.37 (e)). Throughout we will view $CT(f)$ as a subspace of $C(X)$ with the Hausdorff metric and continue to use the Lyapunov function metric on $\mathcal{B}_f$.

Observe that we can now regard a chain component B as either a subset of $\mathcal{C}(f)$, an element of $\mathcal{B}_f$, or an element of $CT(f)$. This fifty percent increase in ambiguity will still be settled by context.

1.6. Summary

In the next lemma we collect some of our observations for future reference. For any $A \subset X$ and a continuous map f on X, we define the *stable set for A* as:

$$W^s_f(A) = \{x : \omega(x,f) \subset A\}. \tag{1.27}$$

When A is an attractor this set is the basin of attraction (see part (c) of the following lemma). If f is a homeomorphism then we also define the *unstable set* for A:

$$W^u_f(A) = W^s_{f^{-1}}(A) = \{x : \alpha(x,f) \subset A\} \tag{1.28}$$

For such f, if R is a repellor then $W^u_f(R)$ is the *repelling basin* for R.

LEMMA 1.5. *Let f be a continuous map on X.*

(a) *If B is a chain component for f then B is a closed, f-invariant subset of X. In particular, if $x \in B$ then $\omega(x,f) \subset B$.*

(b) *For any $x \in X$, $\omega(x,f)$ is a nonempty, closed, invariant set contained in some chain component for f. If the orbit of x meets some inward set U then $\omega(x,f)$ is contained in the associated attractor $A = \omega(U,f)$.*

(c) *If U is an inward set with associated attractor $A = \omega(U,f)$ then*

$$W^s_f(A) = W^s_f(U) = \bigcup_{i=0}^{\infty} f^{-i}(U^\circ) \tag{1.29}$$

is the basin of attraction for A.

(d) *For any $x \in X$, $\omega(x,f)$ is a terminal chain component iff $\omega(x,f) = \omega\mathcal{C}(x,f)$. For $x \in \mathcal{C}(f)$, $\omega(x,f)$ is a terminal chain component iff $\omega(x,f) = \mathcal{C}(x,f)$, in which case x is a recurrent point.*

(e) *Let U be inward for f with associated attractor A and V' be inward for f^{-1} with associated repellor R'. For any $x \in X$, if $\omega(x,f)$ meets $U \cap V'$ then*

$\omega(x, f) \subset A \cap R'$. If f is a homeomorphism then $A \cap R'$ is the largest invariant subset of $U \cap V'$, that is, $A \cap R' = \cap_{n=-\infty}^{+\infty} f^n(U \cap V')$. (Note that here, R' is not necessarily the repellor that is dual to A; in fact in that case the result is rather boring since then $A \cap R' = \emptyset$.)

PROOF. (a) All that is left to show is that $f(B) = B$. Fix a point $b \in B$. By its definition, B is the set of points x such that $x \in \mathcal{C}(b, f)$ and $b \in \mathcal{C}(x, f)$. Since every point of B is chain recurrent, it follows from this characterization and (1.4) that $f(B) \subset B$. The fact that $f(B) = B$ follows from the same argument that established (1.7).

(b) and (c). If y_1, y_2 are both in $\omega(x, f)$ then there are arbitrarily long segments of the forward orbit of x that go from arbitrarily close to y_1 to arbitrarily close to y_2, Consequently $y_1 \in \omega\mathcal{C}(y_2, f)$. This says that any two points of $\omega(x, f)$ are in the same chain component B.

Since the orbit of x meets the inward set U, or sufficiently large j, $f^j(x)$ is in U°, and hence $\omega(f^j(x), f) \subset U$. It is well known, and easy to check, that

$$\omega(x, f) = \omega(f^j(x), x) \tag{1.30}$$

for each positive j, so $\omega(x, f)$ is an invariant set contained in U. However, the attractor $A = \omega(U, f)$ is the largest invariant set in U, so $\omega(x, f) \subset A$. This completes the proof of (b) and equation (1.29). The fact that the basin of attraction of A coincides with $W^s(A)$ now follows from (1.16).

(d) Now suppose $x \in X$ and B is the chain component containing $\omega(x, f)$. If $\omega(x, f)$ is a terminal chain component then $\omega(x, f) = B$ and $\mathcal{C}(B, f) = B$. It follows from (1.6) that $\omega(x, f) = \omega\mathcal{C}(x, f)$. Conversely, with our notation, it is always the case that

$$\omega(x, f) \subset B \subset \omega\mathcal{C}(x, f)$$

(see (1.6) for the second inclusion). Thus, if $\omega(x, f) = \omega\mathcal{C}(x, f)$ then all three of these sets are equal. We have already noted in (1.19) that $\omega\mathcal{C}(x, f)$ is a quasi-attractor, so it follows from Proposition 1.3 that $B = \omega(x, f)$ is terminal. The second sentence of (d) follows from (1.6).

(e) If $\omega(x, f)$ meets $U \cap V'$ then the chain component B containing $\omega(x, f)$ meets $U \cap V'$. Since the only chain recurrent points in U are contained in A, we see that $B \subset A$. Similarly, $B \subset R'$. *A fortiori*, $\omega(x, f) \subset A \cap R'$.

Attractors are always invariant, and for a homeomorphism we know that f-invariance is the same as f^{-1}-invariance, and that R' is an attractor of the map f^{-1}. It follows that $A \cap R'$ is f-invariant. If E is an f-invariant subset of $U \cap V'$ then $E \subset \cap_{n=0}^{\infty} f^n(U) = A$ by f-invariance and $E \subset \cap_{n=0}^{\infty} f^{-n}(V') = R'$ by f^{-1}-invariance. Thus, $E \subset A \cap R'$. Hence, $A \cap R'$ is the largest f-invariant subset of $U \cap V'$, which is clearly $\cap_{-\infty}^{+\infty} f^n(U \cap V')$. $\square$

CHAPTER 2

Periodic Decompositions and Adding Machines

Overview: Several of our proofs are based on the existence of inward sets which are the union of a finite number of pairwise disjoint closed pieces with the map acting essentially as a cyclical permutation on these pieces. We define subsets of X consisting of points that are in such inward sets where the diameter of each piece is arbitrarily small. It will be important to distinguish the case where the number of the pieces is bounded as the diameter goes to 0 from the case where the number goes to infinity – under appropriate hypotheses on X the former cases leads to attracting periodic orbits, while the latter leads to adding machine invariant sets in the closure of the set of periodic orbits.

2.1. Periodic decompositions

For a continuous map f on X and a closed, forward invariant subset A of X, a *k-periodic decomposition* for A is a sequence $\{A_0, A_1, \ldots, A_{k-1}\}$ of pairwise disjoint closed sets, with union A, which satisfy:

$$f(A_{i-1}) \subset A_i \tag{2.1}$$

for $i = 1, 2, \ldots k$ with A_k defined to be A_0. In fact, we extend the sequence $\{A_i\}$ periodically by:

$$A_{i+k} = A_i \qquad i \in \mathbf{Z}, \tag{2.2}$$

after which (2.1) holds for all integers i. For $\epsilon > 0$, the sequence is called an *(ϵ, k)-periodic decomposition* if in addition each A_i has diameter less than ϵ. The A_i's are called the *pieces* of the decomposition.

A closed set A is called a *k-periodic forward invariant set*, or *invariant set* or *inward set* or *attractor*, if it is a forward invariant set, invariant set, etc. which admits a k-periodic decomposition. Similarly, for an *(ϵ, k)-periodic forward invariant set*, *invariant set*, etc. If $\{A_i\}$ is a k-periodic decomposition for a forward invariant set A then it is easy to see that A is invariant iff

$$f(A_{i-1}) = A_i \tag{2.3}$$

for all i, while A is inward iff

$$f(A_{i-1}) \subset\subset A_i \tag{2.4}$$

for all i.

LEMMA 2.1. *Let $f : X \to X$ be continuous.*

(a) *If A, U are nonempty, closed, forward invariant subsets of X with $A \subset U$, and $\{U_i\}$ is an (ϵ, k)-periodic decomposition of U then $\{A \cap U_i\}$ is an (ϵ, k)-periodic decomposition of A. If $A \subset\subset U$ then $A \cap U_i \subset\subset U_i$ for all i. Furthermore, the*

Hausdorff distance $d(A, U)$ *is less than* ϵ. *In particular, this applies if* U *is an inward set and* $A = \omega(U, f)$ *is the associated attractor.*

(b) *If* $\{A_i\}$ *is an* (ϵ, k)*-periodic decomposition for a quasi-attractor* A *and* O *is an open set with* $A \subset O$ *then there exists an inward set* $U \subset O$ *with an* (ϵ, k)*-periodic decomposition* $\{U_i\}$ *satisfying* $A_i = A \cap U_i$ *for all* i. *If* A *is an attractor then* U *can be chosen so that* $A = \omega(U, f)$.

PROOF. (a) Condition (2.1) for $\{U_i\}$ implies the analogous condition for $\{A \cap U_i\}$. $A \cap U_i \subset\subset U_i$ when $A \subset\subset U$ because $U_i^\circ = U_i \cap U^\circ$. Since A is nonempty and forward invariant, for each i $A \cap U_i$ is nonempty. As each U_i has diameter less than ϵ, $U \subset V_\epsilon(A)$ and so $d(A, U) < \epsilon$.

(b) Let ϵ_1 be positive and such that $2\epsilon_1$ is smaller than the differences $\epsilon -$ diameter(A_i) for $i = 0, \ldots, k-1$ and smaller than the distance between distinct pieces of the decomposition for A. Thus, $\{\overline{V_{\epsilon_1}(A_i)} : i = 0, \ldots, k-1\}$ is a sequence of disjoint closed sets of diameter less than ϵ and with union $\overline{V_{\epsilon_1}(A)}$. By shrinking ϵ_1 further we can assume $\overline{V_{\epsilon_1}(A)} \subset O$.

By continuity we can choose a positive ϵ_2 with $\epsilon_2 \le \epsilon_1$ such that $f(\overline{V_{\epsilon_2}(A_{i-1})}) \subset \overline{V_{\epsilon_1}(A_i)}$ for $i = 1, \ldots, k$. Because the $\overline{V_{\epsilon_1}(A_i)}$'s are pairwise disjoint we see that for each i,

$$\overline{V_{\epsilon_2}(A)} \cap \overline{V_{\epsilon_1}(A_i)} = \overline{V_{\epsilon_2}(A_i)}. \tag{2.5}$$

Because A is a quasi-attractor, the inward sets which contain it form a neighborhood base for A (see (1.21)). So we can choose an inward set U such that $A \subset U \subset \overline{V_{\epsilon_2}(A)}$. Let $U_i = U \cap \overline{V_{\epsilon_2}(A_i)}$ for $i = 0, 1, \ldots, k-1$. Clearly, $A_i \subset U_i$ and $\{U_i\}$ is a decomposition of U into disjoint closed sets. Furthermore, since U is forward invariant, $f(U_{i-1}) \subset U \cap f(\overline{V_{\epsilon_2}(A_{i-1})}) \subset U \cap \overline{V_{\epsilon_1}(A_i)}$. Intersecting equation (2.5) with U we see that this last set is U_i. Thus, (2.1) holds for $\{U_i\}$. By part (a), defining $\widetilde{A}_i = U_i \cap A$ gives a decomposition for A. Since $A_i \subset \widetilde{A}_i$ for all i and these sequences have the same union, $A_i = \widetilde{A}_i$ for all i.

If A is an attractor then we can choose U so that $\omega(U, f) = A$, as well (see GTDS Theorem 3.3). $\square$

REMARK 2.1. If U is inward in (a) then each U_i is an inward set for the k-th iterate f^k and $A \cap U_i$ is the associated attractor $\omega(U_i, f^k)$. To see this check that $A \cap U_i = \cap_{n=0}^{\infty} f^n(U) \cap U_i = \cap_{m=0}^{\infty} f^{mk}(U_i) \cap U_i$.

2.2. The sets Π_∞ and $\Pi_{\infty,\infty}$

When f and X as above we define for positive integers k, n:

$$\Pi_{(n,k)}(f) = \bigcup \{U : U \text{ is an } (\frac{1}{n}, k) \text{ -periodic inward set}\} \subset X$$

$$\Pi_\infty(f) = \bigcap_{n=1}^{\infty} \bigcup_{k=1}^{\infty} \Pi_{(n,k)}(f) \tag{2.6}$$

$$\Pi_{\infty,\infty}(f) = \bigcap_{n=1}^{\infty} \bigcap_{j=1}^{\infty} \bigcup_{k=j}^{\infty} \Pi_{(n,k)}(f). \tag{2.7}$$

Thus, $x \in \Pi_\infty(f)$ if it is contained in (ϵ, k)-inward sets for arbitrarily small positive ϵ and $x \in \Pi_{\infty,\infty}(f)$ if, in addition, the period k can be chosen arbitrarily large. The capital Pi is a mnemonic for "periodic inward".

To understand the meaning and importance of these sets, suppose that x is in the intersection of a sequence $\{U^n\}$ where U^n is an (ϵ_n, k_n)-periodic inward set with $\{\epsilon_n\} \to 0$. If the periods $\{k_n\}$ remain bounded then by going to a subsequence we can assume that there exists a constant k such that $k_n = k$ for all n. Then $d(f^k(x), x) \leq \epsilon_n$ for all n and so $f^k(x) = x$. Thus, the orbit of x is periodic and $\{U^n\}$ is a neighborhood base of inward sets for the orbit. So by Proposition 1.3 the periodic orbit is a terminal chain component. When the sequence $\{k_n\}$ is unbounded the orbit of x is infinite but the motion of x appears to be approximately periodic.

2.3. Adding machines

When x is a periodic point for f we will call a positive integer k *the period* of x when $f^k(x) = x$ and the points $x, f(x), \ldots, f^{k-1}(x)$ are distinct. Of course, $f^l(x) = x$ iff the period k divides l, written $k|l$. Thus, the restriction of f to the orbit of x is conjugate to translation by 1 on the finite additive cyclic group of integers mod k, which we will equip with the discrete topology and denote by $\mathbf{Z}_k$. If $\{k_n\}$ is a sequence of positive integers tending to ∞ with $k_n | k_{n+1}$ then $\mathbf{Z}_{k_{n+1}}$ projects to $\mathbf{Z}_{k_n}$ and the inverse limit of the sequence is called the associated *profinite* group. As the inverse limit of a sequence of finite cyclic groups such a group is compact, zero-dimensional and monothetic, i.e. the orbit of 0 under translation by 1 is dense. A continuous map on a compact space which is conjugate to such a translation is called an *adding machine* (see Buescu and Stewart (1995)). An adding machine is a minimal homeomorphism of a very special sort. A point x in X will be called an *adding machine point* if it is recurrent, i.e. $x \in \omega(x, f)$, and the restriction of f to the invariant set $\omega(x, f)$ is an adding machine map. The orbit closure of x is then called an *adding machine subset.*

PROPOSITION 2.2. *Let f be a continuous map on X.*

(a) A point $x \in X$ lies in $\Pi_\infty(f)$ iff it satisfies the conditions (1) *x is either a periodic point or an adding machine point, and* (2) *$\omega(x, f)$ is a terminal chain component for f.*

(b) If $x \in \Pi_\infty(f)$ then x is an adding machine point iff $x \in \Pi_{\infty,\infty}(f)$.

PROOF. Let $x \in \Pi_\infty(f)$. We first show that x is recurrent. Given $\epsilon > 0$, let U be an (ϵ, k)-periodic inward set U with $x \in U$. By Lemma 1.5(b), $\omega(x, f)$ is contained in the attractor associated with U and so by Lemma 2.1(a), the Hausdorff distance $d(\omega(x, f), U) < \epsilon$. Because $x \in U$ and $\epsilon > 0$ was arbitrary we have $x \in \omega(x, f)$.

Now begin with $\epsilon_1 = 1$ and let $\{U_i^1\}$ be an (ϵ_1, k_1)-periodic decomposition of an inward set U^1 containing x. By renumbering if necessary we can assume $x \in U_0^1$. Notice that $f^i(x) \in f^i(U_0^1) \subset\subset U_i^1$ for $i = 1, 2, \ldots$. As above $x \in \omega(x, f)$ and $\{\omega(x, f) \cap U_i^1\}$ is a periodic decomposition for $\omega(x, f)$ with $\omega(x, f) \cap U_i^1 \subset\subset U_i^1$. So we can choose a positive $\epsilon_2 < \epsilon_1/2$ so that $\overline{V_{\epsilon_2}(\omega(x, f) \cap U_i^1)} \subset U_i^1$. Since $x \in \Pi_\infty(f)$ there exists $\{U_i^2\}$ an (ϵ_2, k_2)-periodic decomposition of an inward set U^2 with $x \in U_0^2$. Since $f^i(x) \in \omega(x, f) \cap U_i^2 \cap U_i^1$ we have $U_i^2 \subset V_{\epsilon_2}(\omega(x, f) \cap U_i^1) \subset U_i^1$.

Inductively, we define for $n = 1, 2, \dots$, a constant $\epsilon_{n+1} < \epsilon_n/2$ with

$$\overline{V_{\epsilon_{n+1}}(\omega(x,f) \cap U_i^n)} \subset U_i^n,$$

and then $\{U_i^{n+1}\}$ an $(\epsilon_{n+1}, k_{n+1})$-periodic decomposition of an inward set U^{n+1} with $x \in U_0^{n+1}$. As above $U_i^{n+1} \subset U_i^n$ for all i.

Notice that $U_{k_{n+1}}^{n+1} = U_0^{n+1}$ and these are contained in $U_{k_{n+1}}^n$ and U_0^n respectively. Since $U_i^n \cap U_j^n = \emptyset$ unless $i \equiv j \bmod k_n$ we must have $k_n | k_{n+1}$.

Let $B = \cap_n U^n$ and $\widetilde{B}$ be the inverse limit of cyclic group projections $\mathbf{Z}_{k_{n+1}} \to \mathbf{Z}_{k_n}$. Associate to $y \in B$ the sequence $\{j_n\}$ with j_n the congruence class mod k_n defined by the condition $y \in U_{j_n}^n$. If $\{j_n\}$ is a sequence in $\widetilde{B}$ then $\cap U_{j_n}^n$ is a single point y by the Cantor Intersection Theorem. So the map from B to $\widetilde{B}$ is bijective and continuity is clear. This homeomorphism clearly maps the restriction of f to B to the translation by 1 on the group $\widetilde{B}$. So the restriction of f to B is an adding machine, or a periodic orbit. So B is a minimal $\mathcal{C}(f)$-forward invariant subset of X. It follows from Proposition 1.3 that $B = \omega(x, f)$ is a terminal chain component.

Notice that if $x \in \Pi_{\infty,\infty}(f)$ then the inward sets U^n can be chosen so that $k_{n+1} > k_n$. In that case the group $\widetilde{B}$ is profinite and not finite and B is an adding machine and not a periodic orbit. If $x \notin \Pi_{\infty,\infty}(f)$ then the sequence $\{k_n\}$ remains bounded and so must eventually stabilize. Then $\widetilde{B}$ is a cyclic group and B is a periodic orbit.

For the converse observe that if x is either a periodic point or an adding machine point then $x \in \omega(x, f)$ and for every n, $\omega(x, f)$ admits an $(\frac{1}{n}, k)$-periodic decomposition for some k. If, in addition, $\omega(x, f)$ is a terminal chain component, it is a quasi-attractor and so is contained in some $(\frac{1}{n}, k)$-periodic inward set by Lemma 2.1(b). Thus, $x \in \Pi_{(n,k)}(f)$. Intersecting we see that $x \in \Pi_\infty(f)$. □

PROPOSITION 2.3. *Let A be a chain transitive subset for a continuous map f on X. If A meets $\Pi_\infty(f)$ then A is a periodic orbit or adding machine, terminal chain component contained in $\Pi_\infty(f)$. In particular, A is zero-dimensional.*

If X is locally connected then any zero-dimensional, terminal chain component B is contained in $\Pi_\infty(f)$ and so is a periodic orbit or adding machine set. If, in addition, B is an attractor (rather than just a quasi-attractor) then B is a periodic orbit.

PROOF. If $x \in A \cap \Pi_\infty(f)$ and B is the chain component containing A then $\omega(x, f) \subset A \subset B \subset \mathcal{C}(x, f)$. By Proposition 2.2 x is a recurrent point and $\omega(x, f)$ is a periodic orbit or adding machine terminal chain component. Because distinct chain components are disjoint $\omega(x, f) = B$ and so $A = \omega(x, f)$.

For the converse we assume that B is a zero-dimensional, terminal chain component and that X is locally connected. Given a positive integer n we show that $B \subset \Pi_{(n,k)}(f)$ for some k. We sketch the argument from Hirsch and Hurley (1997), see also Buescu and Stewart (1995) and Melbourne et al. (1993).

Because B is zero-dimensional we can cover B by a sequence of open sets $\{O_1, \dots, O_l\}$ of diameter less than $1/n$, with disjoint closures. By Proposition 1.3 B is a quasi-attractor. So we can choose an inward set U_1 with $B \subset U_1 \subset \cup_j O_j$. Let $\{G_0, \dots, G_{k-1}\}$ list the components of U_1° which meet B. This is a disjoint, and hence finite, open cover of B because $B \subset U_1^\circ$ and X is locally connected. Each G_i is contained in some O_j and so $\overline{G_i}$ has diameter $< 1/n$. Each $f(\overline{G_i})$ meets $f(B) = B$

and is a connected subset of U_1°. So $f(\overline{G_i})$ is contained in some G_s. Thus, $U = \cup_i \overline{G_i}$ is an inward set decomposed into disjoint pieces $\{\overline{G_0}, \dots, \overline{G_{k-1}}\}$ of diameter less than $1/n$. What remains is to show that the sequence can be rearranged to satisfy (2.1). That is, one must show that the map $i \to s$ defined on $\{0, \dots, k-1\}$ by $f(\overline{G_i}) \subset G_s$ consists of a single cycle. The map is clearly onto. To show that this permutation is a single cycle, choose $\delta > 0$ a Lebesgue number for the cover $\{G_i \cap B\}$. Because B is a chain transitive subset we can connect any two points in B by a δ-chain in B, i.e. a sequence $\{x_o, x_1, \dots, x_n\}$ with $d(f(x_j), x_{j+1}) < \delta$. So if $x_j \in G_{i_j}$ then $f(x_j) \in G_{i_{j+1}}$ by the Lebesgue number condition. That is, $i_{j+1} = s(i_j)$. The cyclic property follows.

If B is an attractor then we can choose U_1 so that $\omega(U_1, f) = B$. Since $B \subset U \subset U_1$ we have $\omega(U, f) = B$ as well. For $i = 0, \dots, k-1$ the intersection $B \cap \overline{G_i}$ is $\omega(\overline{G_i}, f^k)$. This set is the intersection of a decreasing sequence of compact connected sets. Hence, it is connected and so is a singleton because B is zero-dimensional. Since B is finite it is a periodic orbit. □

COROLLARY 2.4. *Let f be a continuous map on X and $x \in X$. If $\omega(x, f) \cap \Pi_\infty(f) \neq \emptyset$ then $\omega(x, f) = \omega\mathcal{C}(x, f)$ and this set is a periodic orbit or adding machine, terminal chain component contained in $\Pi_\infty(f)$. In particular,*

$$x \in W_f^s(\Pi_\infty(f)).$$

PROOF. By Lemma 1.5(b), $\omega(x, f)$ is contained in a chain component B and by Proposition 2.3, B is a periodic orbit or adding machine, terminal chain component contained in $\Pi_\infty(f)$. In particular, B is a minimal subset for f and so equals the invariant subset $\omega(x, f)$ of B. Because $\omega(x, f) = B$ is a terminal chain component, Lemma 1.5 (d) implies $\omega\mathcal{C}(x, f) = \omega(x, f)$. □

2.4. Decompositions of $\mathcal{U}$ type

Now assume that $\mathcal{U}$ is a collection of open subsets of X. An (ϵ, k)-periodic decomposition $\{U_i\}$ is said to be *of $\mathcal{U}$ type* if for each i, U_i is the closure of a member of $\mathcal{U}$. An inward set U is called an *(ϵ, k)-periodic inward set of $\mathcal{U}$ type* if it admits such a decomposition. Define:

$$\Pi_{(n,k)}(f, \mathcal{U}) = \bigcup\{U : U \text{ is an } (\frac{1}{n}, k) - \text{periodic inward set of } \mathcal{U} \text{ type}\}. \tag{2.8}$$

and $\Pi_\infty(f, \mathcal{U})$, $\Pi_{\infty,\infty}(f, \mathcal{U})$ by analogy with (2.7), substituting $\Pi_{(n,k)}(f, \mathcal{U})$ for $\Pi_{(n,k)}(f)$ in the latter two definitions. For any $\mathcal{U}$

$$\begin{cases} \Pi_\infty(f, \mathcal{U}) \subset \Pi_\infty(f) \\ \Pi_{\infty,\infty}(f, \mathcal{U}) = \Pi_\infty(f, \mathcal{U}) \cap \Pi_{\infty,\infty}(f). \end{cases} \tag{2.9}$$

The inclusion is clear. The equation then follows because by Proposition 2.2 $\Pi_{\infty,\infty}(f)$ consists of adding machine points while the remaining points of $\Pi_\infty(f)$ are periodic, i.e.,

$$\text{Per}(f) \cap \Pi_\infty(f) = \Pi_\infty(f) \backslash \Pi_{\infty,\infty}(f), \tag{2.10}$$

where $\text{Per}(f)$ denotes the periodic points of f.

LEMMA 2.5. *Assume that $\mathcal{U}$ is a collection of open subsets of X and that f is a continuous map on X.*

$\Pi_\infty(f,\mathcal{U})$ and $\Pi_{\infty,\infty}(f,\mathcal{U})$ are G_δ subsets of X which are unions of terminal chain components. In each case, the associated chain components form a G_δ subset of $\mathcal{B}_f$ and a G_δ subset of $CT(f)$. Furthermore, the stable sets $W^s_f(\Pi_\infty(f,\mathcal{U}))$ and $W^s_f(\Pi_{\infty,\infty}(f,\mathcal{U}))$ are G_δ subsets of X.

PROOF. As the intersections of unions of inward sets, $\Pi_\infty(f,\mathcal{U})$ and $\Pi_{\infty,\infty}(f,\mathcal{U})$ are each $\mathcal{C}(f)$-forward invariant and so each contains any chain component it meets. So each is a union of terminal chain components by Proposition 2.2.

For any inward set U, $U\backslash(U^\circ)$ is in the proper basin of the associated attractor-repellor pair. So from (1.17) we have

$$U \cap \mathcal{C}(f) = U^\circ \cap \mathcal{C}(f). \tag{2.11}$$

So we obtain Π_∞ and $\Pi_{\infty,\infty}$ if in the definitions we replace each U by U°. Thus, Π_∞ and $\Pi_{\infty,\infty}$ are G_δ subsets of X.

Each chain component which meets an inward set U is contained in it and by (2.11) such sets form a clopen subset of $\mathcal{B}_f$. It follows that the chain components in Π_∞ and $\Pi_{\infty,\infty}$ form G_δ subsets of $\mathcal{B}_f$.

For the projection map $\beta : CT(f) \to \mathcal{B}_f$ the preimage of a chain component $B \in \mathcal{B}_f$ consists of B in $CT(f)$ alone exactly when the chain component B is a minimal set. This is true of the Π_∞ chain components and so the set of $\Pi_\infty(f,\mathcal{U})$ chain components in $CT(f)$ is the preimage of the G_δ set of $\Pi_\infty(f,\mathcal{U})$ chain components in $\mathcal{B}_f$, and similarly for $\Pi_{\infty,\infty}(f,\mathcal{U})$. Thus, each is a G_δ subset of $CT(f)$.

By parts (b) and (c) of Lemma 1.5 we see that for positive integers n, k:

$$W^s_f(\Pi_{(n,k)}(f,\mathcal{U})) = \bigcup f^{-i}(U^\circ) \tag{2.12}$$

where the union of all (n,k)-periodic inward set U of type $\mathcal{U}$ and all nonnegative integers i.

This is clearly open and so the result follows from

$$\begin{cases} W^s_f(\Pi_\infty(f,\mathcal{U})) = \bigcap_{n=1}^\infty \bigcup_{k=1}^\infty W^s_f(\Pi_{(n,k)}(f,\mathcal{U})), \\ W^s_f(\Pi_{\infty,\infty}(f,\mathcal{U})) = \bigcap_{n=1}^\infty \bigcap_{j=1}^\infty \bigcup_{k=j}^\infty W^s_f(\Pi_{(n,k)}(f,\mathcal{U})). \end{cases} \tag{2.13}$$

This in turn follows from the observation that if $\{G_\alpha\}$ is a collection of subsets of X then

$$\begin{cases} \mathcal{C}(G_\alpha, f) \subset G_\alpha \text{ for all } \alpha \\ \text{implies} \\ W^s_f(\bigcap_\alpha G_\alpha) = \bigcap_\alpha W^s_f(G_\alpha), \end{cases} \tag{2.14}$$

because $\omega(x,f) \cap G_\alpha \neq \emptyset$ implies $\omega(x,f) \subset G_\alpha$ by chain transitivity of $\omega(x,f)$. $\square$

Let $H(X)$ denote the automorphism group of X, i.e. $h \in H(X)$ if h is a homeomorphism of X onto itself. A collection $\mathcal{U}$ of open sets is called $H(X)$ *invariant* if

$$O \in \mathcal{U} \text{ and } h \in H(X) \Rightarrow h(O) \in \mathcal{U}. \tag{2.15}$$

If $\mathcal{U}$ is $H(X)$ invariant, $h \in H(X)$ and $\{U_i\}$ is a (δ,k)-periodic decomposition of type $\mathcal{U}$ for an inward set for f with δ an ϵ modulus of uniform continuity for f, then

$\{h(U_i)\}$ is in (ϵ, k)-periodic decomposition of type $\mathcal{U}$ for an inward set for hfh^{-1}. So we obtain

$$h(\Pi_\infty(f,\mathcal{U})) = \Pi_\infty(hfh^{-1},\mathcal{U}) \tag{2.16}$$

with the analogous result for $\Pi_{\infty,\infty}$.

In some cases we obtain

$$\begin{cases} \Pi_\infty(f) = \Pi_\infty(f,\mathcal{U}) \\ \Pi_{\infty,\infty}(f) = \Pi_{\infty,\infty}(f,\mathcal{U}). \end{cases} \tag{2.17}$$

Observe that the second equation follows from the first and the equation in (2.9).

LEMMA 2.6. *Assume $\mathcal{U}$ is a collection of open subsets of X. If $\mathcal{U}$ contains a neighborhood base for every closed subset of X then the equations of* (2.17) *hold.*

PROOF. Let $\{U_i\}$ be an (ϵ, k)-periodic decomposition for an inward set U. By assumption we can choose for $i = 0, \ldots, k-1$ elements $G_i \in \mathcal{U}$ such that

$$f(U_{i-1}) \subset\subset G_i \subset\subset U_i. \tag{2.18}$$

Let $V_i = \overline{G_i}$ and extend periodically to obtain a periodic decomposition $\{V_i\}$ for $V = \cup_{i=0}^{k-1} V_i$. From (2.18) we obtain condition (2.4) for V and so $\{V_i\}$ is an (ϵ, k)-periodic decomposition of type $\mathcal{U}$ for the inward set V. Since we have

$$f(U) \subset\subset V \subset\subset U \tag{2.19}$$

it follows that $\omega(V, f) = \omega(U, f)$, i.e. they have the same attractor.

If $x \in \Pi_\infty(f)$ then for every $\epsilon > 0$ there exists k such that x is contained in some (ϵ, k)-periodic inward set U. By Proposition 2.2, x is then in the attractor $\omega(U, f)$ and so x is contained in the (ϵ, k)-periodic inward set of type $\mathcal{U}$ that we constructed above. □

PROPOSITION 2.7. *Let f be a continuous map on X.*

(a) *If $\mathcal{U}$ is the collection of all open subsets of X then $\mathcal{U}$ is an $H(X)$ invariant basis and the equations of* (2.17) *hold.*

(b) *If X is a Cantor set and $\mathcal{U}$ is the collection of all clopen subsets of X then $\mathcal{U}$ is an $H(X)$ invariant basis and the equations of* (2.17) *hold.*

(c) *If X is locally connected and $\mathcal{U}$ is the collection of all connected open subsets of X then $\mathcal{U}$ is an $H(X)$ invariant basis and the equations of* (2.17) *hold.*

PROOF. That $\mathcal{U}$ is an $H(X)$ invariant basis in each case is obvious. In cases (a) and (b) (2.17) follows from Lemma 2.6.

If X is locally connected and $x \in \Pi_\infty(f)$ then for every $\epsilon > 0$ there exists k such that x is contained in an (ϵ, k)-periodic inward set U and so $\omega(x, f) \subset \omega(U, f)$. By Proposition 2.2, x is recurrent and so $x \in \omega(x, f)$. Now follow the argument from Proposition 2.3. We take the finite number of open components of U° which meet $\omega(x, f)$ and order their closures to obtain a periodic decomposition of an inward set $V \subset U$. By definition V is of type $\mathcal{U}$ for $\mathcal{U}$ the family of connected open subsets. Since each component of V is contained in one of the decomposition pieces of U it follows that V is an (ϵ, l)-periodic inward set for some $l \geq k$. Following the proof of

Lemma 2.6 we see that (2.17) holds even though the hypothesis of the lemma does not hold in this case. □

2.5. Periodicity conditions

We will call a space a *ball* when it is homeomorphic to a closed ball in some Euclidean space and a *cone* when it is homeomorphic to the cone on a polyhedron. A space X has the *Fixed Point Property*, we write that X is *FPP*, if every continuous map f on X has a fixed point. If $j : X \to Y$ is continuous then a *retraction* for j is a continuous map $r : Y \to X$ which is a left inverse for j, i.e.

$$r \circ j = 1_X. \tag{2.20}$$

When such an r exists, j is an *embedding*, i.e. a homeomorphism onto the subspace $j(X)$ of Y. When such a pair $\{r, j\}$ exists we say that Y *retracts* to X. If Y is FPP and Y retracts to X then X is FPP because for a continuous map f on X, any fixed point x of jfr lies in $j(X)$ and $r(x)$ is a fixed point of f.

The Brouwer Fixed Point Theorem says that a ball is FPP. Any cone can be embedded in some ball which then retracts onto it. Hence, any cone is FPP. Any FPP set is connected.

For a space X we define $H(X)$-invariant collections of open subsets of X:

$$\begin{cases} \mathcal{U}_B = \{U : U \text{ is open and } \overline{U} \text{ is a ball}\}. \\ \\ \mathcal{U}_C = \{U : U \text{ is open and } \overline{U} \text{ is a cone}\}. \\ \\ \mathcal{U}_F = \{U : U \text{ is open and } \overline{U} \text{ is FPP}\}. \end{cases} \tag{2.21}$$

Clearly,

$$\mathcal{U}_B \subset \mathcal{U}_C \subset \mathcal{U}_F. \tag{2.22}$$

If X is a polyhedron then $\mathcal{U}_C$ is an $H(X)$-invariant basis and if X is a manifold then $\mathcal{U}_B$ is an $H(X)$ invariant basis. However, the equations in (2.17) need not hold for $\mathcal{U}_B$. It is possible to construct a smooth vector field on the three sphere whose time one diffeomorphism has a quasi-attractor fixed point which is not the intersection of inward balls or even inward cones. Constructing such an example is not too difficult, but rather than interrupt the discussion, we omit the details.

The utility of these special classes is illustrated by

PROPOSITION 2.8. *Let $\mathcal{U}_F$ be the family of open subsets of X whose closure has the Fixed Point Property. Let f be a continuous map on X.*

For every $x \in \Pi_\infty(f, \mathcal{U}_F)$ and every $\epsilon > 0$ there exists a periodic point y for f such that

$$d(f^i(x), f^i(y)) \leq \epsilon \quad \text{for } i = 0, 1, \dots. \tag{2.23}$$

In particular, the set $\Pi_\infty(f, \mathcal{U}_F)$ is contained in the closure of the set of periodic points, i.e.

$$\Pi_\infty(f, \mathcal{U}_F) \subset \overline{\mathrm{Per(f)}}. \tag{2.24}$$

PROOF. There exists for some k an (ϵ, k)-periodic decomposition $\{U_i\}$ of type $\mathcal{U}_F$ such that, after renumbering, $x \in U_0$. f^k restricts to a continuous map on the FPP set U_0. So there exists a point $y \in U_0$ such that $f^k(y) = y$. Since $\{f^i(x), f^i(y)\} \subset U_i$ and this set has diameter at most ϵ, (2.23) follows. □

Notice that we did not show that the periodic point y itself lies in $\Pi_\infty(f, \mathcal{U}_F)$. We do not know whether (2.24) remains true when $\mathrm{Per}(f)$ is replaced by the set of periodic points in $\Pi_\infty(f, \mathcal{U}_F)$, $\Pi_\infty(f, \mathcal{U}_F) \backslash \Pi_{\infty,\infty}(f, \mathcal{U}_F)$.

CHAPTER 3

Semicontinuity and Homogeneity

Overview: We review the definitions of upper and lower semicontinuity for set-valued mappings and apply the result that for such maps the set of continuity points is residual. A particular application is in giving sufficient conditions to ensure that the nonwandering set and the chain recurrent set are the same, for all homeomorphisms in a residual subset. In order to describe this result in reasonable generality, we consider actions by groups other than the group of all homeomorphisms on X, and we describe the homogeneity conditions needed to make are arguments go through.

3.1. Semicontinuity

Many of our genericity results are derived by using semicontinuous set valued mappings. Assume Z is a metric space, not necessarily compact, that X is a compact metric space, and let $C(X)$ denote the space of nonempty closed subsets of X, topologized by the Hausdorff metric. A map $F : Z \to C(X)$ can also be viewed as a *pointwise closed relation* $F : Z \to X$, i.e. a subset of $Z \times X$, with the property that $F(z)$ is a closed subset of X for every $z \in Z$. We refer to GTDS for some basic results; in that reference the results are phrased in the language of relations rather than that of set valued mappings.

A map $F : Z \to C(X)$ is called *upper semicontinuous at a point* $z \in Z$, or just, *usc at* z, if for any open subset O of X such that $F(z) \subset O$, $\{w \in Z : F(w) \subset O\}$ is a neighborhood of z in Z. Equivalently, F is usc at z if for any sequence $\{z_n\}$ in Z,

$$\{z_n\} \to z \text{ in } Z \Rightarrow \limsup\{F(z_n)\} \subset F(z), \tag{3.1}$$

where the limsup operator is defined as in (1.1). F is called *lower semicontinuous at a point* z, or *lsc at* z, if for any open subset O of X such that $F(z) \cap O \neq \emptyset$, then $\{w \in Z : F(w) \cap O \neq \emptyset\}$ is a neighborhood of z. Equivalently, F is lsc at z if for any sequence $\{z_n\}$ in Z

$$\{z_n\} \to z \text{ in } Z \Rightarrow F(z) \subset \liminf\{F(z_n)\}, \tag{3.2}$$

where for a sequence $\{Z_n\}$ of closed subsets of X we define

$$\liminf\{Z_n\} \underset{\text{def}}{=} \bigcap_{\epsilon>0} \bigcup_{k} \bigcup_{n \geq k} V_\epsilon(Z_n). \tag{3.3}$$

It is easy to check that the lim inf is closed and is contained in the lim sup, see GTDS page 126. If F is both usc at z and lsc at z, then F is *continuous at* z, in the usual sense that $\{z_n\} \to z$ in Z implies that $d(F(z_n), F(z)) \to 0$, so that

$$\liminf\{F(z_n)\} = \limsup\{F(z_n)\} = F(z). \tag{3.4}$$

F is said to be *usc* (or *lsc*) when it is usc at every point of Z (resp. lsc at every point of Z). Viewed as a relation, F is usc exactly when it is a closed relation (i. e., a closed subset of $Z \times X$). Given a relation $G \subset Z \times X$ and $q \in Z$, we let

$$G(q) = \{x \in X \ : \ (q, x) \in G\}.$$

All of the above equivalences are proved in GTDS as part of Proposition 7.11.

In the following lemma, we employ the language of relations. The reason is that in part (b), given a map $F : X \to C(X)$ it is necessary to distinguish the graph of F (a subset of $Z \times C(X)$) from the subset of $Z \times X$ defined by

$$F' = \{(z, x) \ : \ x \in F(z) \ \}.$$

Of course for each $z \in Z$ the fibers over z can be identified, but the difference is in the topologies: the graph of F being closed, is equivalent to F being continuous as a map between compact metric spaces. On the other hand, F' being closed only means that F is upper semicontinuous as a mapping into $C(X)$. We use relations to try to clarify this distinction.

LEMMA 3.1. *Assume Z is a completely metrizable space and X is a compact metric space. Let $F : Z \to C(X)$.*

(a) *If F is either usc or lsc then the set of continuity points for F is a dense G_δ subset of Z.*

(b) *Assume F is lsc and define $G \subset Z \times X$ to be the closure of F, viewing F as a relation from Z to X (that is, $G = \overline{\{(z, x) : x \in F(z) \ \}}$). G is thus a closed relation, so that G is usc, and $F \subset G$ as relations. If F is continuous at $z \in Z$ then $F(z) = G(z)$. Conversely, if $F(z) = G(z)$ then F and G are continuous at z.*

PROOF. (a) This result from Kuratowski (1932) is proved in Takens (1971) where its usefulness for dynamical systems theory is demonstrated. See also GTDS Theorem 7.19.

(b) GTDS Corollary 7.13 shows that continuity of F at z implies $F(z) = G(z)$. On the other hand, if $z_n \to z$ then by semicontinuity:

$$\begin{array}{rccc} F(z) \subset & \liminf\{F(z_n)\} & \subset & \limsup\{F(z_n)\} \\ & \cap & & \cap \\ & \liminf\{G(z_n)\} & \subset & \limsup\{G(z_n)\} \subset G(z) \end{array} \tag{3.5}$$

so if $F(z) = G(z)$, all of these expressions are equal. □

3.2. Prolongation

The *first prolongation* of the f-orbit of x is defined as

$$\mathcal{P}(x, f) = \bigcap_{\epsilon > 0} \overline{\mathcal{O}(V_\epsilon(x), f)}. \tag{3.6}$$

As an application of part (b) of the lemma, we show that for a residual set of $x \in X$, the first prolongation of X is just the orbit closure of x. The argument is straightforward; lower semicontinuity of the map $x \to \overline{\mathcal{O}(x, f)}$ follows directly from the continuity of f, and it is clear that $\{(x, y) : x \in X, \ y \in \mathcal{P}(x, f)\}$ is the closure in $X \times X$ of $\{(x, y) : x \in X, \ y \in \overline{\mathcal{O}(x, f)}\}$. Thus by the lemma we have:

COROLLARY 3.2. *Let f be a continuous map on a compact metric space X. $\{x \in X : \mathcal{P}(x, f) = \overline{\mathcal{O}(x, f)} \ \}$ is a dense G_δ subset of X.*

REMARK 3.1. It is not hard to check that

$$\mathcal{P}(x, f) = \mathcal{O}(x, f) \cup \limsup_{y \to x} \omega(y, f).$$

Since $\mathcal{C}(x, f) = \mathcal{O}(x, f) \cup \omega\mathcal{C}(x, f)$ (see the discussion preceding (1.7)), it then follows directly that for any $x \in X$

$$\mathcal{C}(x, f) = \mathcal{P}(x, f) \Leftrightarrow \omega\mathcal{C}(x, f) = \limsup_{y \to x} \omega(y, f). \tag{3.7}$$

Note also that $x \in \Omega(f)$ if and only if $x \in \mathcal{P}(f(x), x)$.

3.3. The automorphism group

We now consider the topological structure on $H(X)$, the *automorphism group* of X, consisting of homeomorphisms of X onto itself.

On the space $C(X; X)$ of continuous maps on X the topology of uniform convergence is given by the complete, separable metric

$$d(g_1, g_2) = \sup\{d(g_1(x), g_2(x)) : x \in X\}. \tag{3.8}$$

With the topology induced from the space of maps $H(X)$ is a topological group and d is a right invariant metric. In fact, for $g_1, g_2, f \in C(X; X)$ we have

$$d(g_1, g_2) \geq d(g_1 \circ f, g_2 \circ f) \tag{3.9}$$

with equality if f is surjective.

Because $H(X)$ is usually not a closed subset of $C(X; X)$, the metric d is usually not a complete metric on $H(X)$. We can obtain a complete metric by defining on $H(X)$

$$\bar{d}(g_1, g_2) = \max(d(g_1, g_2), d(g_1^{-1}, g_2^{-1})). \tag{3.10}$$

The metrics d and $\bar{d}$ are topologically equivalent on $H(X)$, i.e. they induce the same topology but they are usually not uniformly equivalent, see, e.g. GTDS Proposition 7.20(b).

Thus, when X is a compact metrizable space the automorphism group $H(X)$ is a Polish group.

A topological space Z is called a *Polish space* when the topology is compatible with a complete separable metric. A topological space Z is called a *Baire space* when the intersection of a countable family of dense open subsets of Z is dense. A Polish space admits a countable base and it is a Baire space by the Baire Category Theorem. For a detailed exposition see Bourbaki (1966) Sections IX.5, IX.6 and for groups Section IX.3.

A *Polish group* is a topological group whose underlying topology is Polish. From a countable base for the neighborhoods of the identity element 1 of a Polish group H one can construct a compatible right-invariant metric d on H, i.e. for $g_1, g_2, f \in H$

$$d(g_1 f, g_2 f) = d(g_1, g_2). \tag{3.11}$$

In particular, with $g_2 = 1$ we obtain

$$d(gf, f) = d(g, 1), \tag{3.12}$$

and so with $f = g^{-1}$:

$$d(1, g^{-1}) = d(g, 1). \tag{3.13}$$

An action of a Polish group H on a space X is a continuous map $e : H \times X \to X$ such that

$$\begin{cases} 1(x) = x \text{ and } (h_1 h_2)(x) = h_1(h_2(x)), \\ \qquad \text{where } h(x) \underset{\text{def}}{=} e(h, x). \end{cases} \tag{3.14}$$

For example, the *evaluation map* $e : H(X) \times X \to X$ is an action of the automorphism group on X. On the other hand, (3.14) implies that when H acts on X we obtain a group homomorphism

$$\begin{cases} e^{\#} : H \longrightarrow H(X) \\ e^{\#}(h)(x) \underset{\text{def}}{=} e(h, x). \end{cases} \tag{3.15}$$

Thus, associated with each $h \in H$ is the homeomorphism $e^{\#}(h)$. This is why we adopted the notation $h(x)$ for $e(h, x)$. Furthermore, it is easy to check that continuity of the homomorphism $e^{\#}$ follows from continuity of e by a compactness argument. Uniform continuity with respect to compatible right invariant metrics, then follows in turn. Conversely, given a continuous homomorphism from H to $H(X)$ we obtain an action of H on X by pulling back the evaluation action of $H(X)$.

In the remainder of this section and in the three sections which follow we will consider fixed action of a Polish group H with right invariant metric d on a space X, $e : H \times X \to X$. Such an action can be regarded as a dynamical system, generalizing the iteration schemes considered earlier. This interpretation has been a consistent theme in the history of topological dynamics from Gottschalk and Hedlund (1955) through Auslander (1988). It is not the viewpoint we will adopt.

Instead, we will regard H as the parameter space for the dynamical systems we will consider. We will regard each individual $h \in H$ as a homeomorphism on X defining a dynamical system on X by iteration as described in Sections 1 and 2. To be precise, it is $e^{\#}(h)$ which is the homeomorphism on X. However, as the notation of (3.14) suggests, we will simply speak of the homeomorphism $h \in H$, meaning the associated homeomorphism $e^{\#}(h)$. Observe that when h_1, and h_2 are close in H then their homeomorphisms are uniformly close. The converse need not be true.

We will think of properties of the H action as aspects of the structure of our underlying space. This is literally true when the action is the evaluation of $H = H(X)$ on X.

Given a dynamical system h in H, a perturbation is an element h_1 in H close to h, so that $h_1 h^{-1}$ is close to 1. Each of our structural assumptions on X, or more precisely on the action, is a demand that the supply of homeomorphisms in neighborhoods of 1 in H be rich enough for various constructions.

The most important example – the one to keep in mind – is the evaluation action of $H = H(X)$. We will refer to this as the $H = H(X)$ *case*. The more general actions arise in the study of this case. For example, suppose that X is a manifold with boundary ∂X nonempty. $H(X)$ acts on X but it acts on ∂X as well by restriction. It will prove much more convenient to use this action directly rather than to consider the restriction homomórphism to $H(\partial X)$. This example illustrates why we will not require that our actions be *effective*, i.e. two different elements h_1 and h_2 of H may represent the same homeomorphism.

A subset A of X is called H*-invariant* if

$$h(A) = A \quad \text{for all } h \in H. \tag{3.16}$$

Observe that $h(A) \subset A$ implies $A \subset h^{-1}(A)$ and so, since H is a group, the concept of H forward invariance reduces to H-invariance. For any subset A_0 of X the smallest H-invariant subset containing A_0 is

$$H(A_0) \underset{\text{def}}{=} \{h(x) : h \in H \text{ and } x \in A_0\}, \tag{3.17}$$

and the smallest closed, H-invariant subset containing A is the closure $\overline{H(A_0)}$. As usual, if A_0 is just a single point, $A_0 = \{x\}$, then we will simply write $H(x)$ instead of $H(\{x\})$.

3.4. Homogeneity conditions

We call points $x, y \in X$, H*-isomorphic points* if $y = h(x)$ for some $h \in H$, i.e. if $y \in H(x)$. In the $H = H(X)$ case, we call x and y *homeomorphic points*. H-isomorphism is clearly an equivalence relation. We will reserve the word "orbit" for the dynamics of individual members of H, referring to the equivalence class $H(x)$ as the H*-iso-class* of x and to $\overline{H(x)}$ as the H*-iso- class closure* of x.

Various homogeneity assumptions can be described using dynamic language for the action of H. The action is called *transitive* when X consists of a single H-iso-class, *topologically transitive* when there exists a dense H-iso-class and *minimal* when every H-iso-class is dense. We require a condition with strong uniform openness condition.

Assume that A is an H invariant subset of X. We say that the action H *is generalized homogeneous on* A if for every $\epsilon > 0$ there exists $\epsilon_1 > 0$ so that if $\{x_1, \dots, x_n\}$ and $\{y_1, \dots, y_n\}$ are two lists of n distinct points in A satisfying $d(x_i, y_i) \le \epsilon_1$, $i = 1, \dots, n$ then there exists $h \in H$ such that $d(h, 1) < \epsilon$ and $h(x_i) = y_i$ $i = 1, \dots, n$. We will then refer to ϵ_1 as an ϵ*-modulus of homogeneity.* Observe that ϵ_1 depends only on ϵ – not on the points in the lists or on their number, n.

We will say that the action H is *generalized homogeneous* when it is generalized homogeneous on X. In the case when $H = H(X)$ will simply say that X is *generalized homogeneous on* A or *generalized homogeneous* (when $A = X$).

A manifold X of dimension greater than 2 is easily seen to be generalized homogeneous on $X \backslash \partial X$, see, e.g., GTDS page 151. With a subtler argument this result extends to dimension 2 as well, see Nitecki and Shub (1975) or Oxtoby (1977) and Akin (2000). When ∂X has dimension at least 2 then X is generalized homogeneous on ∂X as well. Versions of these results go back at least to Oxtoby and Ulam (1941); see also Oxtoby (1977).

A Cantor set is clearly generalized homogeneous, see GTDS page 195. The circle is not generalized homogeneous.

PROPOSITION 3.3. *Assume that the action of H is generalized homogeneous on the subset A of X. A is the disjoint union of a finite number of H-iso-classes each* *of which is open relative to A. In addition, with the topology induced from X, A is a Baire space.*

PROOF. For $x \in X$ we can use the action map e to define the continuous map

$$e_x : H \longrightarrow X$$

$$e_x(h) \underset{\text{def}}{=} e(h,x) = h(x). \tag{3.18}$$

Clearly, the image of e_x is the H-iso-class $H(x)$.

Given $\epsilon > 0$ let $\epsilon_1 > 0$ be an ϵ modulus of generalized homogeneity. For any $h \in H$ and $x \in A$,

$$e_x(V_\epsilon(h)) \supset V_{\epsilon_1}(h(x)) \cap A. \tag{3.19}$$

Since A is H-invariant this implies that for $x \in A$

$$H(x) = A \cap V_{\epsilon_1}(H(x)). \tag{3.20}$$

From compactness of $\overline{A}$ it follows that A is covered by finitely many $V_{\epsilon_1}(H(x))$'s and so A consists of a finite number of $H(x)$'s each open relative to A.

To show that A is a Baire space it suffices to show that $H(x)$ is Baire for $x \in A$. Because H is Polish, it is Baire and (3.19) implies that e_x is an open map. So the result follows from: □

LEMMA 3.4. *Let $g : Z_1 \to Z_2$ be a continuous surjection of topological spaces. Assume that for $U \subset Z_1$*

$$U^\circ \neq \emptyset \Rightarrow (\overline{g(U)})^\circ \neq \emptyset. \tag{3.21}$$

If Z_1 is a Baire space then Z_2 is Baire.

PROOF. Assume that $\{O_n\}$ is a sequence of dense, open subsets of Z_2. Clearly each $g^{-1}(O_n)$ is open in Z_1; if one of these set is not dense in Z_1 then there exists $U \subset Z_1$ open, nonempty and disjoint from $g^{-1}(O_n)$. Since O_n is open in Z_2, $\overline{g(U)}$ is disjoint from O_n. Hence, by (3.21) $(\overline{g(U)})^\circ$ is a nonempty open subset disjoint from O_n, contradicting the density of O_n. Hence, $\{g^{-1}(O_n)\}$ is a sequence of dense open subsets of Z_1. Because Z_1 is Baire, $\bigcap g^{-1}(O_n) = g^{-1}(\cap O_n)$ is dense in Z_1 and so its image, $\cap O_n$, is dense in Z_2. This completes the proof of the lemma. Since an open map satisfies (3.21) the proposition is proved as well. □

REMARK 3.2. If a metric space Y is the image of a Polish space under an open continuous map, then Y is Polish (results of this type go back to Sierpinski (1930, 1956 p. 196); a generalization is Theorem 0 of Ostrovsky (2000)). It follows easily that for x in A, $H(x)$ is a Polish space, which implies that each such $H(x)$, and so A as well, is a G_δ subset of X.

3.5. Showing $\Omega(f) = \mathcal{C}(f)$

Now for our main application of generalized homogeneity. It is stated for relations. For $f \in H$ and $x \in X$ let $\mathcal{P}(x,f)$ denote the first prolongation of the orbit of x, as in (3.6). Define relations $\mathcal{C}''$, $\mathcal{P}'' : H \to X \times X$ by

$$\begin{cases} \mathcal{P}'' = \{\ (f,x,y)\ :\ y \in \mathcal{P}(x,f)\ \}, \\ \mathcal{C}'' = \{\ (f,x,y)\ :\ y \in \mathcal{C}(x,f)\ \}. \end{cases} \tag{3.22}$$

We can think of these as being derived in the obvious way from the graphs of the following maps $H \to C(X \times X)$

$$\begin{cases} \mathcal{P}'(f) = \{\ (x, y)\ :\ y \in \mathcal{P}(x, f)\ \}, \\ \mathcal{C}'(f) = \{\ (x, y)\ :\ y \in \mathcal{C}(x, f)\ \}. \end{cases} \tag{3.23}$$

Recall that the associated invariant sets are the nonwandering set and the chain recurrent set:

$$\begin{cases} \Omega(f) = \{x \mid x \in \mathcal{P}(f(x), f)\ \}, \\ \mathcal{C}(f) = \{x \mid x \in \mathcal{C}(f(x), f)\ \}. \end{cases} \tag{3.24}$$

REMARK 3.3. The map $H \to C(X)$, that sends each $f \in H$ to its chain recurrent set $\mathcal{C}(f)$, is also a usc relation on H (see GTDS Theorem 7.23). In particular,

$$\mathcal{C}(f) = \bigcap_{\epsilon > 0} \overline{\bigcup_{d(f,f_1) \le \epsilon} \mathcal{C}(f_1)}, \tag{3.25}$$

where we let f_1 vary over an ϵ-neighborhood of f in H.

PROPOSITION 3.5. *Assume that D is a dense, H-invariant subset of X, and that the action of H is generalized homogeneous on the D. Then*

$$\{f \in H : \mathcal{P}'(f) = \mathcal{C}'(f)\}$$

is a dense G_δ subset of H. Similarly, the set of $f \in H$ for which $\Omega(f) = \mathcal{C}(f)$ is a dense G_δ.

PROOF. It is easy to check that $f \mapsto \mathcal{C}'(f)$ is usc. Showing $\mathcal{P}'$ is lsc is like the argument in Corollary 3.2. We will use our generalized homogeneity assumption to show that $\mathcal{C}''$ is the closure of $\mathcal{P}''$, so that the Proposition follows directly from Lemma 3.1(b). Since $\mathcal{C}'$ is usc, $\mathcal{C}''$ is closed, so it suffices to show that for $(f, x, y) \in \mathcal{C}''$ and $\epsilon > 0$ there exists $(f_1, x_1, y_1) \in \mathcal{P}''$ ϵ-close to (f, x, y). Choose $\epsilon_1 > 0$ an ϵ modulus of generalized homogeneity. We can find an ϵ_1-chain for f, $\{x_0, x_1, \dots, x_n\}$, with $x_0 = x$ and $x_n = y$ and $n \ge 1$. If $f(x_i) = f(x_j)$ for $0 \le i < j < n$ then we can remove $x_{i+1}, \dots, x_j$ from the chain. Thus, we can assume $\{x_1, \dots, x_n\}$ and $\{f(x_0), \dots, f(x_{n-1})\}$ are lists of distinct points, with $d(f(x_{i-1}), x_i) < \epsilon_1$ for $i = 1, \dots, n$. Since D is dense we can perturb the points $\{x_1, \dots, x_n\}$ if necessary to get $x_i \in D$ for all i. By H-invariance, $f(x_i) \in D$ for all i. Let $h \in H$ with $d(h, 1) < \epsilon$ and $h(f(x_{i-1})) = x_i$ for $i = 1, \dots, n$. Thus, with $f_1 = hf$, $d(f_1, f) < \epsilon$ and $y = x_n = f_1^n(x_0) = f_1^n(x)$. Hence, $(f_1, x, y) \in \mathcal{P}''$.

By Lemma 3.1 $\mathcal{P}'$ is continuous at points of a dense G_δ in H and this set is $\{f : \mathcal{P}'(f) = \mathcal{C}'(f)\}$.

The final assertion follows from the above argument if we take y to be x. □

The following result indicates why the equation $\mathcal{P}'(f) = \mathcal{C}'(f)$ is important.

PROPOSITION 3.6. *Assume f is a continuous map on X satisfying $\mathcal{P}'(f) = \mathcal{C}'(f)$. Then*

$$\{x : \omega(x, f) \text{ is a terminal chain component }\} = \{x : \omega(x, f) = \omega\mathcal{C}(x, f)\}$$

$$= \{x : \omega(x, f) = \limsup_{y \to x} \omega(y, f)\}$$

is a dense G_δ subset of X.

PROOF. Lemma 1.5(d) establishes the first equality. By (3.7), the equation $\mathcal{C}(x, f) = \mathcal{P}(x, f)$ implies $\omega\mathcal{C}(x, f) = \limsup_{y \to x} \omega(y, f)$, so the three descriptions agree when $\mathcal{C}'(f) = \mathcal{P}'(f)$. By Corollary 3.2, the set so described is a dense G_δ. □

At times an extension of Proposition 3.5 is useful. For f a continuous map on X and positive integers k, $l_1, l_2, \ldots, l_k$, define subsets of X^{k+1}

$$(3.26) \qquad \begin{cases} f^{(l_1,\ldots,l_k)} = \{(x_0, x_1, \ldots, x_k) : f^{l_i}(x_{i-1}) = x_i, \quad i = 1, \ldots, k\} \\ \mathcal{P}_k(f) = \overline{\bigcup\{f^{(l_1,\ldots,l_k)} : \ l_1, \ldots, l_k = 1, 2, \ldots\}} \\ \mathcal{C}_k(f) = \{(x_0, \ldots, x_k) : \ x_i \in \mathcal{C}(x_{i-1}, f) \quad i = 1, \ldots, k\}. \end{cases}$$

PROPOSITION 3.7. *Assume that the action of H is generalized homogeneous on a dense subset D of X. $\{f \in H : \mathcal{P}_k(f) = \mathcal{C}_k(f)$ for $k = 1, 2, \ldots\}$, is a dense G_δ subset of H.*

PROOF. Just as in Proposition 3.5, we can regard $\mathcal{P}_k$ and $\mathcal{C}_k$ as relations from H to X^{k+1}. $\mathcal{P}_k$ is lsc. $\mathcal{C}_k$ is usc and from generalized homogeneity we can prove $\mathcal{C}_k$ is the closure of $\mathcal{P}_k$. So by Lemma 3.1(b) $\{f : \mathcal{P}_k(f) = \mathcal{C}_k(f)\}$ is residual for each k. Intersect with $k = 1, 2, \ldots$. □

REMARK 3.4. As before, when X is generalized homogeneous, the analogous result holds for $C(X; X)$.

The arguments in the proof of Proposition 3.5 show that if the action is generalized homogeneous on a dense subset then the relation $f \mapsto \mathcal{C}(f)$ is the closure in $H \times X$ of the relation $f \mapsto \Omega(f)$. For technical reasons we will need a label for this closure in general. So we define the *prolongational nonwandering set*

$$(3.27) \qquad PNW_H(f) = \bigcap_{\epsilon > 0} \overline{\bigcup_{d(f_1, f) \le \epsilon} \Omega(f_1)},$$

again with f_1 and f in H.

As a relation in $H \times X$, PNW_H is usc. We clearly have, for any $f \in H$:

$$(3.28) \qquad \Omega(f) \subset PNW_H(f) \subset \mathcal{C}(f).$$

When f is a continuous map on X we can similarly define the prolongational nonwandering set by varying in $C(X; X)$. That is, we use the right side of (3.27) to define $PNW_C(f)$ with f_1 varying in $C(X; X)$. The analogue of (3.28) holds with PNW_H replaced by PNW_C.

CHAPTER 4

Crushing Arguments

Overview: In this section we are concerned with creating inward sets by perturbing a given homeomorphism. This requires certain hypotheses on the topology of X to allow some freedom in making such perturbations – basically we require a a rich supply of sets where such perturbations exist. These sets we call 'sponges', and we introduce notions of 'sponginess' and 'strict sponginess' to describe when a space has rich supplies of sponges. After we have completed these topological preliminaries, we turn our attention to the consequences of being able to more-or-less freely create inward sets via small perturbation, or 'crushing'. There are three general types of these crushing arguments: 'closing', where periodic inward sets, and consequently attractors, are created; 'trapping', where orbits are forced toward particular attractors; and 'period multiplication', where phenomena of one period are shown to lead to phenomena of higher periods.

Throughout this section H is a Polish group acting on a compact metric space X. For $h \in H$ we use the same symbol h to denote the associated homeomorphism on X. $\mathcal{U}$ is assumed to be an H-invariant basis for X. That is, $\mathcal{U}$ is a basis for the topology of X such that

$$O \in \mathcal{U} \text{ and } h \in H \Rightarrow h(O) \in \mathcal{U}. \tag{4.1}$$

For subsets A, B and C of X we will say that a homeomorphism h on X *crushes A into B* rel C if

$$h(A) \subset\subset B \text{ and } h = 1_X \text{ on } C. \tag{4.2}$$

4.1. Sponges

DEFINITION 4.1. A closed, nonempty subset of K of X will be called a *sponge* for the action of H on X if for every open set U_0 containing K there exists an open set U_1 with $K \subset U_1 \subset\subset U_0$ such that via H, U_1 can be crushed rel $X \backslash U_0$ into any open set which meets K. That is, if U is open and $U \cap K \neq \emptyset$ then there exists $h \in H$ satisfying $h = 1_X$ on $X \backslash U_0$ and $h(U_1) \subset\subset U$. We will call U_1 a *crushing neighborhood* for K in U_0. We will call K an *ϵ-sponge* if the crushing elements $h \in H$ can be chosen so that

$$d(h, 1) < \epsilon \tag{4.3}$$

using the metric in H.

LEMMA 4.2. *For every $\epsilon > 0$, there exists $\delta > 0$ such that any δ-sponge has diameter less than ϵ in X. In the $H = H(X)$ case, any sponge with diameter less than ϵ is an ϵ-sponge.*

PROOF. Choose $0 < \epsilon_1 < \epsilon$ and δ an ϵ_1 modulus of continuity for the homomorphism $e^{\#} : H \to H(X)$, so that $d(h, 1) \leq \delta$ implies that the homeomorphism h

satisfies $d(h, 1_X) < \epsilon_1$ with respect to the metric (3.8) on $H(X)$. Suppose that K is a sponge with $x_1, x_2 \in K$ and $d(x_1, x_2) \geq \epsilon$. There is an open set U containing x_2 such that $d(x_1, y) > \epsilon_1$ for all $y \in U$. Because K is a sponge there exists $h \in H$ such that $h(K) \subset U$. In particular, $h(x_1) \in U$ and so $d(x_1, h(x_1)) > \epsilon_1$. Hence, $d(h, 1) > \delta$ in H, and K is not a δ-sponge.

In the $H = H(X)$ case if K has diameter less than ϵ then we can shrink U_0 if necessary so that it, too, has diameter less than ϵ. If $h = 1_X$ on $X \backslash U_0$ then $h(U_0) = U_0$ and so $d(h, 1_X) < \epsilon$, with respect to the metric (3.8) on $H(X)$. □

Here is the prototype of our results using sponges.

PROPOSITION 4.3. *Let $x \in X$ be a periodic point of period k for the homeomorphism $f \in H$. Assume that for every $\epsilon > 0$, x is contained in some ϵ-sponge for the action. Let O_0 be an open subset of X with $x \in O_0$ and let $\epsilon > 0$.*

There exists $h \in H$ and $U \subset X$ such that:

(1) $h = 1_X$ on $X \backslash O_0$ and $d(h, 1) < \epsilon$.

(2) U is an inward set for $g = hf$ with an (ϵ, k)-periodic decomposition of $\mathcal{U}$ type $\{U_i\}$ satisfying $x \in U_0$ and $U_0 \subset O_0$.

PROOF. By shrinking O_0 if necessary, we can assume that

$$f^i(x) \in X \backslash \overline{O_0} \text{ for } 0 < i < k. \tag{4.4}$$

By Lemma 4.2 and the hypothesis we can choose an ϵ-sponge K for the action so that

$$x \in K \text{ and } K \subset O_0. \tag{4.5}$$

Let O_1 be a crushing neighborhood of K in O_0.

Because k is the period of x, we can choose γ so that $\epsilon/2 > \gamma > 0$ and

$$\begin{cases} V_\gamma(x) \subset O_1 \\ V_\gamma(f^i(x)) \subset X \backslash \overline{O_0} \text{ for } 0 < i < k \end{cases} \tag{4.6}$$

and $\{V_\gamma(f^i(x)) : 0 \leq i < k\}$ are pairwise disjoint.

Choose U_k a closed neighborhood of $x = f^k(x)$ with $U_k \subset V_\gamma(x)$. By downward induction choose U_i for $i = k-1, \dots, 0$ to be the closure of an element of $\mathcal{U}$ containing $f^i(x)$ such that $U_i \subset V_\gamma(f^i(x)) \cap f^{-1}(U_{i+1}^\circ)$. The sequence of closed sets $\{U_0, \dots, U_k\}$ satisfies:

$$\begin{cases} x \in U_0^\circ \\ U_0 \cup U_k \subset O_1 \\ U_1 \cup \ldots \cup U_{k-1} \subset X \backslash \overline{O_0} \\ f(U_i) \subset\subset U_{i+1} \text{ for } i = 0, \dots, k-1 \\ \{U_0, \dots, U_{k-1}\} \text{ are pairwise disjoint} \\ \text{diameter } U_i < \epsilon \textit{ for } i = 1, \dots, k-1. \end{cases} \tag{4.7}$$

By definition of the sponge K and crushing neighborhood O_1, there exists $h \in H$ such that $h = 1_X$ on $X \backslash O_0$, $d(h, 1) < \epsilon$ and

$$h(O_1) \subset U_0^\circ. \tag{4.8}$$

With $g = hf$ it is clear that

$$(4.9)\qquad \begin{cases} g = f \text{ on } U_0 \cup \ldots \cup U_{k-2} \\ g(U_{k-1}) \subset h(U_k) \subset h(O_1) \subset U_0^\circ. \end{cases}$$

Consequently, $\{U_0, \ldots, U_{k-1}\}$ is an (ϵ, k) periodic decomposition of $\mathcal{U}$ type for a g-inward set. □

REMARK 4.1. This proof illustrates a subtlety in the definition of a sponge. If we only assume that h crushes K into U rel $X \backslash U_0$ then there exists an open neighborhood U_1 of K such that $h(U_1) \subset\subset U$. However, U_1 would then depend on U. We need to be able to choose the crushing neighborhood U_1 first.

DEFINITION 4.4. The action of H on X is called *spongy* if for every $\epsilon > 0$, open subset O_0 of X and point $y \in O_0$ there exists an open subset O_1 of X satisfying:

(1) $y \in O_1$ and $O_1 \subset O_0$.
(2) For every pair x_1, x_2 of H-isomorphic points in O_1, there exists an ϵ-sponge K such that $\{x_1, x_2\} \subset K \subset O_0$ and K is perfect (i.e. has no isolated points).

The action is called *strictly spongy* if condition (2) is replaced by:

(2)′ For every pair x_1, x_2 of H-isomorphic points in O_1 and every finite subset F of $O_0 \backslash \{x_1, x_2\}$ there exists a perfect ϵ-sponge K such that $\{x_1, x_2\} \subset K \subset O_0 \backslash F$.

As usual, in the $H = H(X)$ case we will call X spongy or strictly spongy.

Notice that in the definition we do not assume that x_1 and x_2 are distinct points. So every point x in O_1 is contained in some ϵ-sponge in O_0.

LEMMA 4.5. *The action is spongy iff for every $\epsilon > 0$ there exists $\delta > 0$ such that any pair $\{x_1, x_2\}$ of H-isomorphic points with $d(x_1, x_2) < \delta$ is contained in some perfect ϵ-sponge K. The action is strictly spongy if, in addition, whenever F is a finite subset of X disjoint from $\{x_1, x_2\}$, K can be chosen disjoint from F.*

PROOF. Assume the action is spongy. Every y in X is contained in some open set O_y such that every H-isomorphic pair in O_y is contained in a perfect ϵ-sponge. Let $\delta > 0$ be a Lebesgue number for the cover $\{O_y : y \in X\}$. So if $d(x_1, x_2) < \delta$ then $\{x_1, x_2\} \subset O_y$ for some y. So if x_1 and x_2 are H-isomorphic they are contained in some perfect sponge.

Conversely, assume the condition of the lemma. Let O_0 be open and $y \in O_0$. Choose $\epsilon > 0$ so that $V_{2\epsilon}(y) \subset O_0$. Use Lemma 4.1 to choose $\epsilon_1 > 0$ with $\epsilon_1 \leq \epsilon$ and such that any ϵ_1-sponge has diameter less than ϵ. Now choose $\delta > 0$ with $\delta \leq \epsilon_1$ so that any H-isomorphic pair with $d(x_1, x_2) < 2\delta$ is contained in some perfect ϵ_1-sponge. Let $O_1 = V_\delta(y)$. Any H-isomorphic pair x_1, x_2 in O_1 satisfies $d(x_1, x_2) < 2\delta$ and so is contained in some perfect ϵ_1-sponge K. Since the diameter of K is at most ϵ, $K \subset O_0$.

We leave to the reader the easy adjustments for the strictly spongy case. □

PROPOSITION 4.6. (a) *Any 1-dimensional manifold X with $\partial X = \emptyset$, i.e. any disjoint union of circles, is spongy.*

(b) *Any 2 dimensional manifold X is spongy and if $\partial X = \emptyset$ then X is strictly spongy.*

(c) *Any manifold X of dimension greater than 2 is strictly spongy.*

(d) *A Cantor set is strictly spongy.*

PROOF. For (a) - (c) observe that if X is a ball of positive dimension in a Euclidean space then any arc K which is p.l. embedded either in ∂X or in $X \backslash \partial X$ is a perfect sponge.

If X is a Cantor set then a nonempty closed subset K of X is a sponge exactly when it is not clopen. For if K is clopen then $U_0 = K$ is a neighborhood of K and so $K \subset U_1 \subset U_0$ implies $U_1 = K = U_0$. A homeomorphism h which equals 1_X on $X \backslash U_0$ must satisfy $h(U_0) = U_0$ and so in this case $h(U_1)$ cannot be a proper subset of U_0. On the other hand, if K is closed but not clopen and U_0 is clopen with $K \subset U_0$ then $U_0 \backslash K \neq \emptyset$. So we can choose a clopen set U_1 such that $K \subset U_1 \subset U_0$ with $U_1 \backslash U_0 \neq \emptyset$. If U is open and meets K we can shrink it to assume $U \subset U_1$ and U is clopen. Because all Cantor sets are homeomorphic we can define $h = 1_X$ on $X \backslash U_0$ so that $h(U_1) = U$ and $h(U_0 \backslash U_1) = U_0 \backslash U$. This is the required crushing homeomorphism.

Since we are in the $H = H(X)$ case we can apply Lemma 4.2 and obtain ϵ-sponges by choosing sponges of diameter less than ϵ. ☐

Now we consider properties of spongy actions.

LEMMA 4.7. *Assume that K is a sponge for the action.*

(a) *If $f \in H$ then $f(K)$ is a sponge for the action. Furthermore, for every $\epsilon > 0$ and $f \in H$ there exists $\delta > 0$ such that $f(K)$ is an ϵ-sponge if K is a δ-sponge.*

(b) *If U_1 is any crushing neighborhood of K and $x \in U_1$ then*

$$K \subset \overline{H(x)}. \tag{4.10}$$

(c) *If A is a connected subset of X and $K \subset A$ then K is connected.*

(d) *If A is an H-invariant subset of X with only finitely many components and $K \subset A$ then K is connected.*

PROOF. (a) If V_0 is a neighborhood of $f(K)$ and U_1 is a crushing neighborhood for K in $f^{-1}(V_0) = U_0$ then $V_1 = f(U_1)$ is a crushing neighborhood for $f(K)$ in V_0. If $V \cap f(K) \neq \emptyset$ and h is a element crushing U_1 into $U = f^{-1}(V)$ rel $X \backslash U_0$ then fhf^{-1} crushes V_1 into V rel $X \backslash V_0$. Since the inner automorphism $h \mapsto fhf^{-1}$ is continuous on H, there exists $\delta > 0$ such that $d(h, 1) < \delta$ implies $d(fhf^{-1}, 1) < \epsilon$.

(b) If U is any open set meeting K there exists $h \in H$ such that $h(U_1) \subset U$. In particular, $H(x) \cap U \neq \emptyset$.

(c) Suppose that K is the disjoint union of two closed nonempty set K^+ and K^-. Choose disjoint open neighborhoods $U_0^\pm$ of $K^\pm$. Choose U_1 a crushing neighborhood for K contained in $U_0 = U_0^+ \cup U_0^-$. With $U = U_0^+$ there exists $h \in H$ such that $h(U_1) \subset U_0^+$ and $h = 1_X$ on $X \backslash U_0$. Because h is the identity on the boundary of U_0 we have

$$U_0^- \cap h^{-1}(U_0^+) = \overline{U_0^-} \cap h^{-1}(\overline{U_0^+}) \tag{4.11}$$

and so this set is clopen. It contains K^-, and so is nonempty, and it is disjoint from K^+. So no connected subset of X contains K.

(d) By (c) it suffices to prove that K is contained in a single component of A. Suppose that $x \in K$ lies in component A_1 of A. Since the components are closed relative to A and there are only finitely many of them, A_1 is open relative to A. Hence there exists U open in X with $x \in U$ and $U \cap A \subset A_1$. Choose $h \in H$ such that $h(K) \subset U$. Because $h(A) = A$, h permutes the components of A. But $h(K) \subset U$ and $K \subset A$ implies h maps every component of A which meets K to A_1. Hence, K meets only one component of A. □

PROPOSITION 4.8. *Assume that the action of H on X is spongy.*

(a) *If X_0 is any closed, H invariant subset of X then the restriction of the action of H to X_0 is spongy. Furthermore, if the action of H on X is strictly spongy then the restricted action on X_0 is strictly spongy.*

(b) *For every $x \in X$, the H-iso-class $H(x)$ is a Baire space with no isolated points, with respect to the topology induced from X.*

(c) *The following conditions on X and the H action are equivalent:*

(1) *X is locally connected.*

(2) *X has only finitely many components.*

(3) *Every sponge in X is connected.*

PROOF. (a) If K is a sponge in X and $x \in K \cap X_0$ then by Lemma 4.7(b) $K \subset \overline{H(x)} \subset X_0$. If K is an ϵ-sponge for the action of H on X then it is clearly an ϵ-sponge for the action of H on X_0. If U_1 is a crushing neighborhood for K in X then $U_1 \cap X_0$ is a crushing neighborhood for K in X_0.

(b) If K is a perfect sponge with $x \in K$ then by Lemma 4.7(b) again, $K \subset \overline{H(x)}$. Since K is perfect and $x \in K$, x cannot be an isolated point of $H(x)$. Now, as in Proposition 3.3, we will consider the continuous surjection $e_x : H \to H(x)$ of (3.18) and then apply Lemma 3.4. This will imply that $H(x)$ is Baire once we check that e_x satisfies condition (3.21).

Let $U \subset H$ satisfy $U^\circ \neq \emptyset$. We can choose $\epsilon > 0$ and $f \in H$ such that $V_\epsilon(f) \subset U$. Use Lemma 4.5 to choose $\delta > 0$ so that every pair in $H(x)$ of diameter less than δ is contained in an ϵ-sponge. Of course, $f(x) \in H(x)$. Assume that $x_1 \in H(x) \cap V_\delta(f(x))$. There is an ϵ-sponge K such that $\{f(x), x_1\} \subset K$. Let O be an arbitrarily small open set containing x_1. There exists a crushing element $h \in V_\epsilon(1_X)$ such that $h(K) \subset O$. In particular, $hf(x) \in O$. It follows that $x_1 \in \overline{e_x(V_\epsilon(f))}$. Since x_1 was arbitrarily chosen we have

$$H(x) \cap V_\delta(f(x)) \subset \overline{e_x(U)}. \tag{4.12}$$

Hence, $f(x)$ is in the $H(x)$ interior of $\overline{e_x(U)}$. This proves condition (3.21).

(c) (1) $\Rightarrow$ (2): The components of a locally connected space are clopen. So a compact, locally connected space has only finitely many components.

(2) $\Rightarrow$ (3) Apply Lemma 4.7 (d).

(3) $\Rightarrow$ (1): Given $\epsilon > 0$ we can apply Lemmas 4.2 and 4.5 to choose a positive $\delta < \epsilon/2$ so that any pair x_1, x_2 of H isomorphic points with $d(x_1, x_2) < 2\delta$ are contained in some sponge of diameter less than $\epsilon/2$. For $y \in X$ we will show that $\overline{V_\epsilon(y)}$ contains a connected neighborhood of y. As y and ϵ are arbitrary this will prove local connectedness.

First choose a sponge K with $y \in K$ and $K \subset V_\delta(y)$. Let U be a crushing neighborhood of K for $V_\delta(y)$.

Fix $x \in U$. For any pair of points x_1, x_2 in $U \cap H(x)$ we can choose a sponge $K(x_1, x_2)$ of diameter less than $\epsilon/2$ and which contains $\{x_1, x_2\}$. Hence $K(x_1, x_2) \subset \overline{V_\epsilon(y)}$.

By assumption (3) each $K(x_1, x_2)$ is connected and so it is easy to see that

$$C(x) = \overline{\bigcup \{K(x_1, x_2) : x_1, x_2 \in U \cap H(x)\}} \tag{4.13}$$

is a closed connected subset of $\overline{V_\epsilon(y)}$. Notice that $K(x_1, x_2) \cap K(x_2, x_3)$ contains x_2.

By Lemma 4.7 (b), $K \subset \overline{H(x)}$ and so, since U is a neighborhood of K, $K \subset \overline{U \cap H(x)}$. Since $x_1, x_2 \in K(x_1, x_2)$ it follows that $\overline{U \cap H(x)} \subset C(x)$ and so K is contained in each of the connected sets $C(x)$. Of course, $x \in C(x)$.

Now vary x:

$$U \subset \bigcup \{C(x) : x \in U\} \subset \overline{V_\epsilon(y)}. \tag{4.14}$$

The union is connected since any two $C(x)$'s meet in K. □

There is an obvious result worth noting because it is false. Suppose that H acts on X and H_0 is a closed subgroup of H. If K is an ϵ-sponge for the restricted H_0 action then it is, of course, an ϵ-sponge for the H action. Nonetheless, it can happen that the H_0 action is spongy while the H action is not. The problem occurs because the H-isomorphism classes might be larger than the H_0-isomorphism classes.

For example, let X consist of three 2 dimensional discs joined along their common boundary circle. Think of two hemispheres with a flat equatorial disc. If H_0 is the group of homeomorphisms which preserve each separate disc then the action is spongy. Once we allow permutation of the discs the action fails to be spongy.

4.2. Crushing

We now describe techniques, which we call *crushing arguments*, to be applied later. These are the major uses of the spongy assumptions. They are best thought of as variations on the theme introduced in Proposition 4.3.

PROPOSITION 4.9 (Closing). *Assume that the action of H on X is spongy. Let $f \in H$ and $\epsilon > 0$. Assume that O_0 is an open subset of X with $O_0 \cap \Omega(f) \neq \emptyset$. There exists $h \in H$, $U \subset X$ and a positive integer k such that:*

(1) *$h = 1_X$ on $X \backslash O_0$ and $d(h, 1_X) < \epsilon$.*
(2) *U is an periodic inward set for $g = hf$ with an (ϵ, k)-periodic decomposition of $\mathcal{U}$ type $\{U_0, \dots, U_{k-1}\}$ satisfying $U_0 \subset O_0$.*

PROPOSITION 4.10 (Trapping). *Assume that the action of H on X is spongy. Let $f \in H$ and $\epsilon > 0$. Assume that $x \in X$ and O_0, O are open subsets of X with $\omega(x, f) \cap O_0 \neq \emptyset$ and $\omega(x, f) \subset O$. There exist $h \in H$, $U \subset X$ and a positive integer k such that:*

(1) *$h = 1_X$ on $X \backslash O_0$ and $d(h, 1_X) < \epsilon$.*
(2) *U is an inward set for $g = hf$ with an (ϵ, k)-periodic decomposition of $\mathcal{U}$ type $\{U_0, \dots, U_{k-1}\}$ satisfying $U_0 \subset O_0$ and $U \subset O$.*
(3) *For the associated attractor $A = \omega(U, g)$, x is in the proper basin of attraction, $W^s_g(A) \backslash A$.*

PROPOSITION 4.11 (Period Multiplication). *Assume that the action of H on X is strictly spongy. Let $f \in H$ and $\epsilon > 0$. Assume that $x \in X$ and O_0 is an open subset of X with $\omega(x, f) \cap O_0 \neq \emptyset$. Assume that W is an inward set for f with k-periodic decomposition $\{W_0, \ldots, W_{k-1}\}$ such that $O_0 \subset W_0$.*

There exist $h \in H$, $U \subset X$ and a positive integer m such that

(1) *$h = 1_X$ on $X \backslash O_0$ and $d(h, 1_X) < \epsilon$.*
(2) *U is an inward set for $g = hf$ with an $(\epsilon, 2mk)$-periodic decomposition of $\mathcal{U}$ type $\{U_0, \ldots, U_{2mk-1}\}$ satisfying $U_0 \cup U_{mk} \subset O_0$ and $U \subset W$.*
(3) *For the associated attractor $A = \omega(U, g)$, x is in the proper basin of the attractor $W_g^s(A) \backslash A$.*

Now we turn to the proofs. Each construction is a sequence of choices and definitions as in Proposition 4.3. Observe that for $f \in H$ all of the points $f^i(x)$ lie in the H iso-class $H(x)$.

PROOF. (Closing) Choose O_1 open, $y \in X$ so that $y \in O_1 \cap \Omega(f)$, $O_1 \subset\subset O_0$ and any two points in O_1 which are H-homeomorphic are contained in some ϵ-sponge in O_0.

Because y is nonwandering $O_1 \cap f^{k_0}(O_1) \neq \emptyset$ for some positive integer k_0. Choose x and a positive integer k_0 so that $x, f^{k_0}(x) \in O_1$. Choose K an ϵ-sponge so that $\{x, f^{k_0}(x)\} \subset K$ and $K \subset O_0$. Define k to be the smallest positive integer such that $f^k(x) \in K$. Thus, $x, f^k(x) \in K$ but $f^i(x) \notin K$ for $0 < i < k$. Choose O_2 open so that

$$(4.15) \qquad \begin{cases} \{x, f^k(x)\} \subset K \subset O_2 \subset\subset O_0 \\ f^i(x) \in X \backslash \overline{O_2} \ \text{ for } \ 0 < i < k. \end{cases}$$

Because K is a sponge we can choose O_3 an open crushing neighborhood of K in O_2, i.e. $K \subset O_3 \subset\subset O_2$ and O_3 can be crushed rel $X \backslash O_2$ into any open set meeting K.

By definition of k, the points $\{f^i(x) : 0 \leq i < k\}$ are distinct and so we can choose a positive $\gamma < \epsilon/2$ such that

$$(4.16) \qquad \begin{cases} V_\gamma(x) \cup V_\gamma(f^k(x)) \subset O_3 \\ V_\gamma(f^i(x)) \subset X \backslash \overline{O_2} \quad 0 < i < k, \end{cases}$$

and $\{V_\gamma(f^i(x)) : 0 \leq i < k\}$ are pairwise disjoint.

Choose U_k a closed neighborhood of $f^k(x)$ with $U_k \subset V_\gamma(f^k(x))$. By downward induction, choose U_i for $i = k-1, \ldots, 0$, the closure of an element of $\mathcal{U}$ containing $f^i(x)$ such that $U_i \subset V_\gamma(f^i(x)) \cap f^{-1}(U_{i+1}^\circ)$. The sequence $\{U_0, \ldots, U_k\}$ of closed sets satisfies:

$$(4.17) \qquad \begin{cases} x \in U_0^\circ \\ U_0 \cup U_k \subset O_3 \\ U_1 \cup \ldots \cup U_{k-1} \subset X \backslash \overline{O_2} \\ f(U_{i-1}) \subset\subset U_i \quad i = 1, \ldots, k \\ \{U_0, \ldots, U_{k-1}\} \text{ are pairwise disjoint} \\ \text{diameter } U_i < \epsilon \quad i = 0, \ldots, k. \end{cases}$$

Choose O_4 any open subset of U_0 with $O_4 \cap K \neq \emptyset$. Choose $h \in H$ so that $h = 1_X$ on $X \backslash O_2$, $d(h,1) < \epsilon$ and $h(O_3) \subset\subset O_4$. Observe that with $g = hf$:

$$\begin{cases} g = f \text{ on } U_0 \cup \ldots \cup U_{k-2} \\ g(U_{k-1}) = h(f(U_{k-1})) \subset h(U_k) \subset\subset O_4 \subset U_0. \end{cases} \tag{4.18}$$

Thus, $\{U_0, \ldots, U_{k-1}\}$ is an (ϵ, k)-periodic decomposition for $U = \cup_{i=0}^{k-1} U_i$, an inward set for g. Furthermore, $U_0 \subset O_3 \subset O_0$ and h is the identity on $X \backslash O_2 \supset X \backslash O_0$. □

Proof. (Trapping) The orbit of x eventually enters and remains in O since $\omega(x,f) \subset O$ and O is open. Define $k_0 = \min\{i \geq 0 : f^j(x) \in O \text{ for all } j \geq i\}$.

Choose $y \in \omega(x,f) \cap O_0$. If $y = f^i(x)$ for some i then $\omega(x,f) \subset O$ implies that $k_0 \leq i$. Hence, we can shrink O_0 if necessary to exclude the points $f^j(x)$ with $0 \leq j < k_0$ since none of these points equals y. Thus, we can assume

$$\begin{cases} y \in O_0 \subset\subset O \\ f^j(x) \in X \backslash \overline{O_0} \quad 0 \leq j < k_0. \end{cases} \tag{4.19}$$

Choose O_1 open so that $y \in O_1$, $O_1 \subset\subset O_0$ and any two points in O_1 which are H-isomorphic in X are contained in some perfect sponge in O_0.

Because $y \in \omega(x,f) \cap O_1$ the orbit of x frequently enters O_1 so we can choose a perfect sponge $K \subset O_0$ such that $f^i(x) \in K$ for at least two distinct positive values of i.

Define k_1 to be the smallest nonnegative integer such that $f^{k_1}(x) \in K$ and k to be the smallest positive integer such that $f^{k_1+k}(x) \in K$. By (4.19) we see that $k_1 \geq k_0$ because $K \subset O_0$. Hence, $f^{k_1+i}(x) \in O$ for all $i \geq 0$.

Choose O_2 open so that

$$\begin{cases} \{f^{k_1}(x), f^{k_1+k}(x)\} \subset K \subset O_2 \subset\subset O_0 \\ f^j(x) \in X \backslash \overline{O_2} \quad 0 \leq j < k_1 \text{ and } k_1 < j < k_1 + k. \end{cases} \tag{4.20}$$

Choose O_3 an open crushing neighborhood of K in O_2. All of the points $\{f^{k_1}(x), \ldots, f^{k_1+k-1}(x)\}$ are distinct and so we can choose a positive $\gamma < \epsilon/2$ such that

$$\begin{cases} V_\gamma(f^{k_1}(x)) \cup V_\gamma(f^{k_1+k}(x)) \subset O_3, \\ V_\gamma(f^{k_1+i}(x)) \subset O \backslash \overline{O_2} \quad 0 < i < k, \end{cases} \tag{4.21}$$

and $\{V_\gamma(f^{k_1+i}(x)) : 0 \leq i < k\}$ are pairwise disjoint.

Choose, as in the previous proof, a sequence $\{U_0, \ldots, U_k\}$ closures of elements of $\mathcal{U}$ satisfying:

$$\begin{cases} f^{k_1}(x) \in U_0^\circ \\ U_0 \cup U_k \subset O_3 \\ U_1 \cup \ldots \cup U_{k-1} \subset O \backslash \overline{O_2} \\ f(U_{i-1}) \subset\subset U_i \quad i = 1, \ldots, k \\ \{U_0, \ldots, U_{k-1}\} \text{ are pairwise disjoint} \\ \operatorname{diam} U_i < \epsilon \quad i = 0, \ldots, k. \end{cases} \tag{4.22}$$

Because K is perfect we can choose O_4 an open subset of U_0 with $O_4 \cap K \neq \emptyset$ and

$$\{x, f^{-1}(x), \ldots, f^{-k}(x)\} \cap O_4 = \emptyset. \tag{4.23}$$

Choose $h \in H$ so that $h = 1_X$ on $X \backslash O_2$, $d(h, 1) < \epsilon$ and $h(O_3) \subset\subset O_4$. The inclusions of (4.18) are again true and so U is an (ϵ, k)-periodic inward set with decomposition $\{U_0, \ldots, U_{k-1}\}$. By (4.22) we also have $U \subset O$. Thus, conditions (1) and (2) follow.

Because $h = 1_X$ on $X \backslash O_2$, (4.20) implies that:

$$\begin{cases} f^j(x) = g^j(x) \quad 0 \leq j < k_1 \\ \\ g^{k_1}(x) = h(f^{k_1}(x)). \end{cases} \tag{4.24}$$

This last is a point of $h(K) \subset U_0$. Thus, the g-orbit of x enters the inward set U which implies $x \in W^s_g(A)$ where $A = \omega(U, g)$. However, from (4.18) we see that $g^k(U_i) \subset f^i(O_4)$ for $i = 0, \ldots, k-1$. Hence, $A \subset g^k(U) \subset \cup_{i=0}^{k-1} f^i(O_4)$. So by (4.23), $x \notin A$. □

Proof. (Period Multiplication) Since $\{W_i\}$ is a k-periodic decomposition for W with O_0 in W_0, we have for $y \in \omega(x, f) \cap O_0$,

$$\begin{cases} y \in O_0 \subset\subset W_0 \\ \\ f^i(W_0) \subset\subset W_i \subset X \backslash \overline{O_0} \quad 0 < i < k. \end{cases} \tag{4.25}$$

Choose O_1 open so that $y \in O_1$, $O_1 \subset\subset O_0$ and any two H-isomorphic points in O_1 are contained in some perfect $(\epsilon/3)$-sponge in O_0 which can be chosen to avoid a finite set.

Because $y \in \omega(x, f) \cap O_1$, the orbit of x enters O_1. Let k_1 be the first entrance time. Because $O_1 \subset W_0$ and $\{W_i\}$ is a k-periodic decomposition the return times of the orbit to O_1 all differ from k_1 by a multiple of k. Let $k_1 + mk$ be the first return time. Hence,

$$\begin{cases} f^{k_1}(x), \ f^{k_1+mk}(x) \in O_1 \\ \\ f^i(x) \notin O_1 \quad i \in \{0, \ldots, k_1 + mk\} \backslash \{k_1, k_1 + mk\}. \end{cases} \tag{4.26}$$

Let $z = f^{k_1}(x)$ and define

$$F_0 = \{f^i(x) : 0 \leq i \leq k_1 + mk\}. \tag{4.27}$$

In what we will call the *simple case*, this set consists of $k_1 + mk + 1$ distinct points. In the remaining case, which we will call the *repeating case*, $f^{i_1}(x) = f^{i_2}(x)$ for some i_1, i_2 with $0 \leq i_1 < i_2 \leq k_1 + mk$. Applying f $k_1 + mk - i_2$ times we see that $f^{k_1+mk}(x) = f^i(x)$ for some i with $0 \leq i < k_1 + mk$. By (4.26) it must be that $i = k_1$ and so x is a periodic point with period mk. In this case, F_0 is the periodic orbit of z, or equivalently, the entire orbit of x. In particular, $x = f^j(z)$ for some j with $0 \leq j < mk$. Notice that in the simple case, x might still be a periodic point, but one with a larger period.

Choose $\delta > 0$ so that

$$V_\delta(z) \cup V_\delta(f^{mk}(z)) \subset O_1, \tag{4.28}$$

and the sets in $\{V_\delta(y) : y \in F_0\}$ are pairwise disjoint. In particular, the mk sets $V_\delta(f^j(z))$, $j = 0, \dots, mk-1$, are disjoint. Define:

$$O_2 \underset{\text{def}}{=} \bigcap_{j=0}^{mk} f^{-j}(V_\delta(f^j(z))). \tag{4.29}$$

Clearly, $z \in O_2$ and O_2 is an open subset of O_1.

By Proposition 4.8(b), $H(x)$ has no isolated points and so, since $z \in H(x) \cap O_2$ we can choose a point

$$z' \in O_2 \cap H(x) \backslash (\bigcup_{j=0}^{mk} f^{-j}(F_0)). \tag{4.30}$$

So by the choice of δ and the definition of O_2 the points $f^j(z')$ for $0 \le j < mk$ are distinct and if we define

$$F_0' = \{f^j(z') : 0 \le j \le mk\}, \tag{4.31}$$

we have

$$F_0 \cap F_0' = \emptyset. \tag{4.32}$$

Choose a point z''

$$z'' \in O_2 \cap H(x) \backslash (F_0 \cup F_0'), \tag{4.33}$$

and define the finite sets:

$$\begin{cases} F = F_0 \cup F_0' \cup \{z''\} \\ F_1 = F \backslash \{f^{mk}(z'), z''\} \\ F_2 = F \backslash \{f^{mk}(z), z'\} \\ F_3 = F \backslash \{z'', z\}. \end{cases} \tag{4.34}$$

We can choose $(\epsilon/3)$-sponges K_1, K_2, K_3 and open sets T_1, T_2, T_3 such that

$$\begin{cases} \{f^{mk}(z'), z''\} \subset K_1 \subset T_1 \subset\subset O_0 \backslash F_1 \\ \{f^{mk}(z), z'\} \subset K_2 \subset T_2 \subset\subset O_0 \backslash F_2 \\ \{z'', z\} \subset K_3 \subset T_3 \subset\subset O_0 \backslash F_3 \end{cases} \tag{4.35}$$

We will use these sponges to move $f^{mk}(z)$ to z', $f^{mk}(z')$ to z'' and z'' to z. The z orbit when it returns to O_1 after mk steps would then land on z'. After mk steps it would then be back at z. We need the interpolation point z'' because otherwise the period doubling adjustment would fail if it should happen that $f^{mk}(z') = z'$. Of course, using sponges we can't make these moves precisely, but we don't require such precision because we are constructing a periodic inward set not a periodic orbit. See Figure 4.1.

Now choose crushing neighborhoods G_α for K_α in T_α ($\alpha = 1, 2, 3$). Choose a positive $\gamma < \min(\delta, \epsilon/2)$ so that $\{V_\gamma(y) : y \in F\}$ are pairwise disjoint and, in addition

$$\begin{cases} V_\gamma(K_\alpha) \subset G_\alpha, \\ V_\gamma(F_\alpha) \cap T_\alpha = \emptyset \quad (\alpha = 1, 2, 3). \end{cases} \tag{4.36}$$

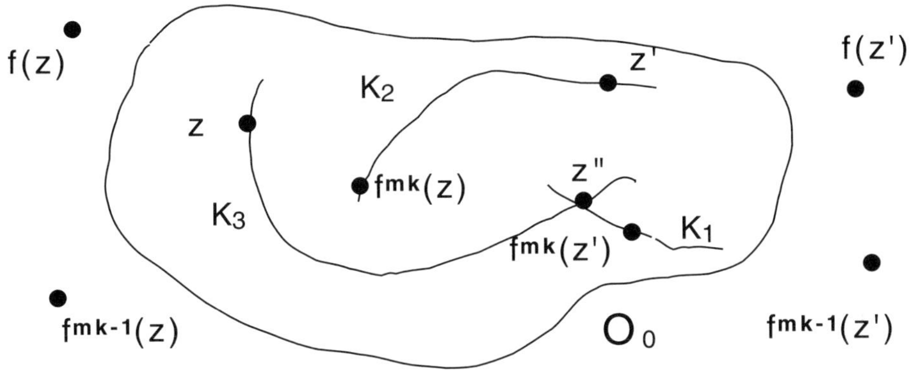

FIGURE 4.1

Since $z, z' \in O_0$ an open subset of W_0 we can shrink γ if necessary to get

$$V_\gamma(f^j(z)) \cup V_\gamma(f^j(z')) \subset W_j \tag{4.37}$$

for $j = 0, \ldots, mk$ where $\{W_j\}$ is the periodic extension of the decomposition for W (see (2.2)).

Now, as usual, we can proceed by downward induction to construct U_j and U'_j for $j = mk, \ldots, 0$, the closures of elements of $\mathcal{U}$ which contain $f(z)$ and $f^j(z')$ respectively and so that:

$$\begin{cases} U_j \subset V_\gamma(f^j(z)),\ U'_j \subset V_\gamma(f^j(z'))\ 0 \le j \le mk \\ f(U_j) \subset\subset U_{j+1},\ f(U'_j) \subset\subset U'_{j+1}\ 0 \le j < mk. \end{cases} \tag{4.38}$$

It follows from the choice of γ that

$$\begin{cases} U_{mk} \subset G_2,\ U'_{mk} \subset G_1 \\ \operatorname{diam} U_j, U'_j < \epsilon \quad 0 \le j \le mk \\ U_j \cup U'_j \subset W_j \quad 0 \le j \le mk. \end{cases} \tag{4.39}$$

Furthermore, the sets $U_0, \ldots, U_{mk-1}, U'_0, \ldots, U'_{mk-1}$ are pairwise disjoint.

Choose open sets O_3, O'_3

$$\begin{cases} O_3 \cap K_3 \neq \emptyset,\ \ O_3 \subset U_0 \\ O'_3 \cap K_2 \neq \emptyset,\ \ O'_3 \subset U'_0 \\ (O_3 \cup O'_3) \cap \{f^{-j}(x) : j = 0, \ldots, mk\} = \emptyset. \end{cases} \tag{4.40}$$

Now we define our crushing elements $h_1, h_2, h_3 \in V_{\epsilon/3}(1_X)$ in H such that

$$(4.41)\qquad \begin{cases} h_\alpha = 1_X \text{ on } X \backslash T_\alpha \ (\alpha = 1, 2, 3) \\ h_1(G_1) \subset\subset V_\gamma(z'') \subset G_1 \cap G_3 \\ h_2(G_2) \subset\subset O_3' \subset U_0' \\ h_3(G_3) \subset\subset O_3 \subset U_0. \end{cases}$$

Define $h = h_3 h_2 h_1$ and $g = hf$.

First, observe that by (4.36)

$$(4.42)\qquad h = 1_X \text{ on } V_\gamma(F_1 \cap F_2 \cap F_3).$$

Since $f^j(z), f^j(z') \in F_1 \cap F_2 \cap F_3$ for $0 < j < mk$ it follows that h is the identity on $U_j \cup U_j'$ for $0 < j < mk$. Consequently,

$$(4.43)\qquad g = f \text{ on } U_j \cup U_j' \quad 0 \leq j < mk - 1.$$

Also, $f(U_{mk-1}) \subset U_{mk} \subset V_\gamma(f^{mk}(z)) \subset G_2$. Notice that

$$(4.44)\qquad V_\gamma(f^{mk}(z)) \cap T_1 = \emptyset$$

because $f^{mk}(z) \in F_1$. Hence, h_1 is the identity on U_{mk}, or as we will say, h_1 ignores U_{mk}. Then h_2 maps G_2 into $O_3' \subset U_0' \subset V_\gamma(z')$. Since $z' \in F_3$, h_3 ignores this set. Hence,

$$(4.45)\qquad g(U_{mk-1}) \subset\subset O_3' \subset U_0'.$$

Similarly, $f(U'_{mk-1}) \subset U'_{mk} \subset V_\gamma(f^{mk}(z')) \subset G_1$ and h_1 maps the latter set into $V_\gamma(z'')$ which is ignored by h_2. Then h_3 maps this subset of G_3 to O_3. Thus,

$$(4.46)\qquad g(U'_{mk-1}) \subset\subset O_3 \subset U_0.$$

Combining these inclusions with (4.39) and (4.38) we see that, after renumbering, $\{U_0, \ldots, U_{mk-1}, U_0', \ldots, U'_{mk-1}\}$ is an $(\epsilon, 2mk)$ periodic decomposition of $\mathcal{U}$ type for an inward set U for g with $U \subset W$. This proves (1) and (2) of Proposition 4.11.

Finally, by (4.43)

$$(4.47)\qquad \bigcup_{j=0}^{mk-1} (g^j(O_3 \cup O_3')) = \bigcup_{j=0}^{mk-1} (f^j(O_3 \cup O_3')).$$

The attractor $A = \omega(U, g)$ is contained in this set and by (4.40) the point x is not. So x is not in A. To complete the proof we must show that x is in the basin of the attractor.

In what we called above the simple case $\{f^i(x) : 0 \leq i < k_1\} \subset F_1 \cap F_2 \cap F_3$ and so

$$(4.48)\qquad \begin{cases} g^i(x) = f^i(x) \quad 0 \leq i < k_1 \\ g^{k_1}(x) = hf^{k_1}(x) = h(z). \end{cases}$$

In this simple case, $z \neq f^{mk}(z)$ and $z \in F_1 \cap F_2$. Hence, $h(z) = h_3(z) \in O_3 \subset U_0$. So x is in $W_g^s(A)$.

In the repeating case, $x = f^j(z)$ for some $0 \leq j < mk$. If $j < mk - 1$ then by (4.43), $x = g^j(z)$ and so x is in the basin of A because $z \in U_0$. Finally, if $x = f^{mk-1}(z)$ then $g(x) = hf^{mk}(z)$. By (4.44), $h_1 f^{mk}(z) = f^{mk}(z)$ and

$$(4.49)\qquad g(x) = hf^{mk}(z) = h_3 h_2 f^{mk}(z) = h_2 f^{mk}(z) \in O_3' \subset U_0'.$$

Thus in this case, too, x is in the basin of A. □

CHAPTER 5

Topological Horseshoes

Overview: Our goal here is to show how to make perturbations that create invariant sets that are semiconjugate to a full two shift. As usual, there are some topological preliminaries, then we give conditions sufficient to ensure the existence of such a semiconjugacy. Finally we describe how to create these conditions by small perturbations of a given map.

5.1. Connected successions of subsets

For our next crushing argument we require a few topological definitions. A *partition* of X is a finite, pairwise disjoint cover of X by nonempty, closed subsets. Thus, each member of the partition is clopen. For example, the components of a locally connected space form a partition of the space.

A pair of subsets A_1 and A_2 of X are called *separated in* X if there exists a partition of X into a pair of sets (X_1, X_2) which extends the pair (A_1, A_2), i.e. $X = X_1 \cup X_2$. $X_1 \cap X_2 = \emptyset$ and $A_\alpha \subset X_\alpha = \overline{X_\alpha}$ $(\alpha = 1, 2)$. Otherwise we say that the subsets A_1 and A_2 are *not separated from each other in* X. Note that these notions have nothing to do with the individual topological properties of A_1 and A_2, but depend only on the way these sets are situated in X. As examples, if $A_1 = \emptyset$ or $A_2 = \emptyset$ then X clearly separates A_1 and A_2; if X is a connected space then no pair of nonempty subsets is separated in X.

A sequence $\{A_1, \ldots A_k\}$ of subsets of X is called a *succession* of subsets if

$$A_i \cap A_j \neq \emptyset \Leftrightarrow |i - j| \leq 1, \quad 1 \leq i, j \leq k. \tag{5.1}$$

A *connected succession* of subsets is a succession $\{A_1, \ldots A_k\}$ of closed subsets such that $A \equiv \cup_{\alpha=1}^{k} A_\alpha$ does not separate A_1 and A_k. In general, for i, j with $1 \leq i \leq j \leq k$ we define $A(i, j) = \cup_{\alpha=i}^{j} A_\alpha$. So $A(i, i) = A_i$ and $A(1, k) = A$.

What we are calling "successions" have in other places been called "chains". In view of the prominence of chain recurrence in this work, we thought it best to try to avoid possible confusion caused by the similar names, and so we do not use the standard name here.

LEMMA 5.1. *Let $f : X \to Y$ be a continuous map.*

(a) (Extension) *If X does not separate A_1 and A_2 and $f(A_\alpha) \subset B_\alpha$ $(\alpha = 1, 2)$ then Y does not separate B_1 and B_2.*

(b) (Excision) *If X is the union of closed subsets $A_1, A_2, \widetilde{X}$ with $A_1 \cap A_2 = \emptyset$ then A_1 and A_2 are separated in X iff $A_1 \cap \widetilde{X}$ and $A_2 \cap \widetilde{X}$ are separated in $\widetilde{X}$.*

(c) *If $\{A_1, \ldots, A_k\}$ is a connected succession of subsets of X and i, j are integers with $1 \leq i, j \leq k$ and $i + 1 < j$ then neither X nor $A(i, j)$ separates A_i and A_j. Furthermore, $A_i \cap A(i + 1, j - 1)$ and $A_j \cap A(i + 1, j - 1)$ are not separated in $A(i + 1, j - 1)$.*

(d) *Assume* $\{B_1, \dots B_k\}$ *is a connected succession of subsets of* Y *with* $B = \cup_{i=1}^k B_i$. *Assume* A *is a closed subset of* X *with* $f(A) \subset B$. *Define* $A_i = A \cap f^{-1}(B_i)$, $i = 1, \dots, k$. *If* A_1 *and* A_k *are not separated in* A *then* $\{A_1, \dots, A_k\}$ *is a connected succession of subsets of* X *with union* A.

PROOF. (a) If (Y_1, Y_2) is a partition of Y separating (B_1, B_2) then with $X_\alpha = f^{-1}(Y_\alpha)$ ($\alpha = 1, 2$), (X_1, X_2) is a partition of X separating (A_1, A_2).

(b) If $(\widetilde{X}_1, \widetilde{X}_2)$ is a partition of $\widetilde{X}$ separating $(A_1 \cap \widetilde{X}, A_2 \cap \widetilde{X})$ then with $X_\alpha = \widetilde{X}_\alpha \cup A_\alpha$ ($\alpha = 1, 2$), (X_1, X_2) is a partition of X separating (A_1, A_2). For the reverse implication apply the Extension Property (a) to the inclusion map of $\widetilde{X}$ into X.

(c) Since A_1 and A_k are not separated in A, $A(1, i)$ and $A(j, k)$ are not separated in A by the Extension Property applied to the identity map on A. By the succession condition (5.1):

$$A(1, i) \cap A(i + 1, j - 1) = A_i \cap A(i + 1, j - 1)$$

$$A(j, k) \cap A(i + 1, j - 1) = A_j \cap A(i + 1, j - 1).$$

So by Excision, these two sets are not separated in $A(i+1, j-1)$. Apply Extension to the inclusion map of $A(i + 1, j - 1)$ into $A(i, j)$ and into A, to get that A_i and A_j are not separated, either in $A(i, j)$ or in A.

(d) Clearly, $A_i \cap A_j \subset f^{-1}(B_i \cap B_j)$ is empty if $|i - j| > 1$ by (5.1) applied to the B succession. If $i \leq j$ and $|i - j| \leq 1$ but $A_i \cap A_j = \emptyset$ then $(A(1, i), A(i + 1, k))$ is a partition of A extending (A_1, A_k), contradicting the assumption that A_1 and A_k are not separated in A. So $\{A_1, \dots, A_k\}$ is a connected succession of subsets of X. □

5.2. Topological horseshoes

Let f be a continuous map on X. An inward set U for f is called a *topological horseshoe* if there is a connected succession of subsets $\{U_1, \dots, U_5\}$ with $U = \cup_{\alpha=1}^5 U_\alpha$ such that

$$(5.2) \qquad \begin{cases} f(U_1) \cup f(U_5) \subset\subset U_1 \\ f(U_3) \subset\subset U_5. \end{cases}$$

Since U is an inward set we also have

$$(5.3) \qquad f(U) \subset\subset U.$$

In such a case we will say that U is a topological horseshoe for f *with respect to the succession* $\{U_1, \dots, U_5\}$.

Because of the strong inclusions there exists $\epsilon > 0$ so that if g is a continuous map on X with $d(f, g) \leq \epsilon$, the inclusions of (5.2) and (5.3) still hold when f is replaced by g. Thus, U is a topological horseshoe, with the same succession of subsets, for any continuous map close enough to f.

Notice that if X_0 is a closed f-invariant subset for X and U is a topological horseshoe for f with respect to the succession $\{U_1, \dots, U_5\}$ then $U \cap X_0$ is a topological horseshoe for the restriction of f to X_0, with respect to $\{U_1 \cap X_0, \dots, U_5 \cap X_0\}$ iff the latter is a connected succession of subsets of X_0. By Lemma 5.1(d) this is true iff $U_1 \cap X_0$ and $U_5 \cap X_0$ are not separated in $U \cap X_0$.

Let Σ^+ and Σ denote the one-sided and two-sided symbol spaces, $\Sigma^+ = \{0,1\}^{\mathbf{N}}$ and $\Sigma = \{0,1\}^{\mathbf{Z}}$ where $\mathbf{Z}$ and $\mathbf{N}$ are the sets of integers and nonnegative integers respectively. Equipped with the product topology these are Cantor spaces. The shift maps s on Σ^+ and Σ are defined by

$$\sigma(a)(i) = a(i+1). \tag{5.4}$$

The map σ is a homeomorphism on Σ but only a surjection on Σ^+. Both are topologically transitive maps. A continuous map f on a space X is called *topologically transitive* when the subset

$$\text{Trans}(f) = \{x \in X : \omega(x,f) = X\} \tag{5.5}$$

is nonempty. In that case, our standing assumption that X is compact metric implies that Trans(f) is a dense G_δ that is f-invariant. A proof of this standard result can be found in GTDS, Theorem 4.12.

PROPOSITION 5.2. *Let f be a continuous map on X and U be a topological horseshoe for f. There exist closed subsets B, B_0 of X and a continuous map $\pi^+ : B \to \Sigma^+$ such that*

(1) *$B_0 \subset B \subset U$ with B a chain component for f and B_0 an f-invariant set such that the restriction of f to B_0 is topologically transitive. B is contained in the attractor $\omega(U,f)$ associated with U.*

(2) *The restriction to B_0 of map π^+ is surjective and π^+ maps f on B to σ on Σ^+, i.e.*

$$\pi^+ \circ f(x) = \sigma \circ \pi^+(x) \quad x \in B. \tag{5.6}$$

If f is a homeomorphism on X then a continuous map $\pi : B \to \Sigma$ can be chosen so that condition (5.6) *above still holds with π and Σ replacing π^+ and Σ^+.*

PROOF. Let $\{U_1, U_2, U_3, U_4, U_5\}$ be a connected succession of closed sets with union U, satisfying (5.2). Define $E_0 = U_2$ and $E_1 = U_4$. These are disjoint closed sets by the succession condition (5.1). Define the closed subset E of U and the map $\pi^+ : E \to \Sigma^+$ by

$$\begin{cases} E = \bigcap_{i=0}^{\infty} f^{-i}(E_0 \cup E_1) \\ \pi^+(x)(i) = \epsilon \Leftrightarrow f^i(x) \in E_\epsilon \quad \text{for } \epsilon = 0 \text{ or } 1. \end{cases} \tag{5.7}$$

E is the closed, forward invariant subset consisting of all $x \in U$ such that $f^i(x) \in E_0 \cup E_1$ for $i = 0, 1, \ldots$. Since $E_0 \cap E_1 = \emptyset$ the coding sequence $\pi^+(x)$ is well-defined. Furthermore, $f^i(f(x)) \in E_{\pi^+(f(x))(i)}$ while $f^{i+1}(x) \in E_{\pi^+(x)(i+1)}$. Equation (5.6) follows for all x in E.

With $w = (w_0, \ldots, w_{n-1})$ a word of length n (each $w_i = 0$ or 1), we define

$$E_w = \bigcap_{i=0}^{n-1} f^{-i}(E_{w_i}). \tag{5.8}$$

Each E_w is a closed subset of U and $\{E_w \cap E\}$ is a partition of E into 2^n disjoint closed subsets as w varies over the words of length n. In fact,

$$x \in E_w \cap E \Leftrightarrow \pi^+(x)(i) = w_i \quad 0 \le i < n. \tag{5.9}$$

This proves that π^+ is continuous. To show that π^+ is surjective it suffices to show that each E_w is nonempty. Then if $a \in \Sigma^+$ and a^n is the word of length

n forming the initial segment of a, $(\pi^+)^{-1}(a)$ is the intersection of the decreasing sequence of nonempty compact sets $\{E_{a^n}\}$.

We prove by induction on the length n of the word w that

$$(5.10)\qquad \begin{cases} f^n(E_w) \subset U \text{ and} \\ E_w \cap f^{-n}(U_1) \text{ and } E_w \cap f^{-n}(U_5) \text{ are not separated in } E_w. \end{cases}$$

In particular, it follows that these sets are nonempty.

Initial Step: By (5.2) and the definitions of E_0 and E_1 we have

$$(5.11)\qquad \begin{cases} U_2 \cap U_1 \subset E_0 \cap f^{-1}(U_1) \text{ and } U_2 \cap U_3 \subset E_0 \cap f^{-1}(U_5) \\ U_4 \cap U_5 \subset E_1 \cap f^{-1}(U_1) \text{ and } U_4 \cap U_3 \subset E_1 \cap f^{-1}(U_5) \end{cases}$$

By part (c) of Lemma 5.1, $U_1 \cap U_2$ and $U_3 \cap U_2$ are not separated in $U_2 = E_0$. By the inclusions of (5.1) and the Extension Property $E_0 \cap f^{-1}(U_1)$ and $E_0 \cap f^{-1}(U_5)$ are not separated in E_0. The argument for E_1 is similar. Clearly, $f(E_0) \cup f(E_1) \subset f(U) \subset U$.

Inductive Step: Assume that (5.10) holds for the length n word w. We complete the inductive step by showing that it holds for the two length $n+1$ words $w0$ and $w1$.

The inductive hypothesis allows us to apply part (d) of Lemma 5.1 to the map $f^n : E_w \to U$ to define $E_w^i = E_w \cap f^{-n}(U_i)$ $(i = 1, 2, 3, 4, 5)$ obtaining a connected succession of 5 sets with union E_w. Observe that $E_{w0} = E_w^2$ and $E_{w1} = E_w^4$. In particular, $f^{n+1}(E_{w0} \cup E_{w1}) \subset f(U_2 \cup U_4) \subset U$.

Applying part (c) of Lemma 5.1 to the succession on E_w we see that $E_w^2 \cap E_w^1$ and $E_w^2 \cap E_w^3$ are not separated in E_w^2. This says that $E_w \cap f^{-n}(U_2 \cap U_1)$ and $E_w \cap f^{-n}(U_2 \cap U_3)$ are not separated in E_{w0}. By the first two inclusions of (5.11) these two sets are contained in $E_{w0} \cap f^{-(n+1)}(U_1)$ and $E_{w0} \cap f^{-(n+1)}(U_5)$ and so the latter are not separated in E_{w0} by Extension. The argument for E_{w1} is similar and completes the inductive step.

Thus, E is a closed, forward invariant subset of U with $\pi^+(E) = \Sigma^+$. By the usual Zorn's Lemma argument we can choose B_0 a closed, forward invariant subset of E with $\pi^+(B_0) = \Sigma^+$ and which is minimal with respect to these properties. Because π^+ is topologically transitive on Σ^+ there exists $a \in \Sigma^+$ such that $\omega(a, \sigma) = \Sigma^+$. Because σ is surjective there exists $x \in B_0$ such that $\pi^+(x) = a$. Since π^+ maps f to σ on E, compactness implies that $\pi^+(\omega(x, f)) = \omega(a, \sigma) = \Sigma^+$. But $\omega(x, f)$ is a closed forward invariant subset of B_0 and so by minimality $\omega(x, f) = B_0$. Hence, the restriction of f to B_0 is topologically transitive and B_0 is invariant. Hence, B_0 is contained in the largest invariant subset of U, the attractor $A = \omega(u, f)$. $B_0 = \omega(x, f)$ is contained in some chain component B. B meets U and so $B \subset A$ as well.

Now let $\widetilde{U} = U_1 \cup U_3 \cup U_5$. By (5.2) this is an inward set as is U_1. Furthermore, $f^2(\widetilde{U}) \subset U_1$ and so $\omega(\widetilde{U}, f) = \omega(U_1, f)$, the attractor associated with U_1 which we will denote A_1.

If $y \in A_1 \cap E$ then for all i, $f^i(y) \in U_1 \cap (E_0 \cup E_1) \subset E_0$ and so $\pi^+(y)(i) = 0$ for all i. It follows that the chain component B is disjoint from $\widetilde{U}$. For if $B \cap \widetilde{U} \neq \emptyset$ then $B \subset \omega(\widetilde{U}, f) = A_1$, and so, in particular, $x \in B \cap E$ would be in $A_1 \cap E$. This is impossible since $\pi^+(x)$ is a transitive point for σ.

Thus, $B \subset U\backslash\widetilde{U} \subset E_0 \cup E_1$. Because B is invariant it follows that $B \subset E$. Hence the map π^+ is defined on B as required.

If $\widetilde{U}'$ is $X\backslash\widetilde{U}^\circ$ with associated repellor $\widetilde{R}$, then $B \subset A \cap \widetilde{R}$; s Now if f is a homeomorphism then f^{-1} is defined on B and B_0 because these are invariant sets. So we can define the continuous map π on B by $\widetilde{\pi}(x) = \{\pi^+(f^{-i}(x)) : i = 0, 1, \ldots\}$ mapping B into the inverse limit $\Sigma^+ \overset{\sigma}{\leftarrow} \Sigma^+ \overset{\sigma}{\leftarrow} \Sigma^+ \ldots$. $\widetilde{\pi}$ is surjective on B by compactness because each $\pi^+ \circ f^{-i}$ is surjective on B. Similarly, if $\pi_0 : \Sigma \to \Sigma^+$ is the "forgetful" map then π_0 maps σ on Σ to σ on Σ^+ and $\widetilde{\pi}_0$ maps Σ homeomorphically onto the same inverse limit. Define $\pi = \widetilde{\pi}_0^{-1} \circ \widetilde{\pi} : B \to \Sigma$. $\square$

5.3. Perturbing to a horseshoe

Now we return to the context of an action of a Polish group H on the space X. We want to construct topological horseshoes for strictly spongy actions. This will require some connectedness conditions, as one might expect. In addition a technical difficulty comes up in the proof which will require us to specialize to the case where H is $H(X)$ or at least is a closed subgroup of $H(X)$. The problem arises in a construction which we will now examine.

LEMMA 5.3. *Assume the action of H on X is strictly spongy. Given B a closed, connected subset of X, O and O' open subsets of X and points x in O and $x' \in O'$. Assume that*

$$(5.12) \qquad \begin{cases} x \in B \subset O \cap \overline{H(x)} \\ x' \in B \cap H(x) \cap O'. \end{cases}$$

Let F be a finite subset of X disjoint from $\{x, x'\}$. Let $\epsilon > 0$.

(a) *There exists $\{A_1, \ldots, A_m\}$ a succession of sponges such that $x \in A_1$, $x' \in A_m$ and $A_j \cap H(x) \neq \emptyset$ for $j = 1, \ldots, m$, and open sets W_j for $j = 1, \ldots, m$ such that*

$$(5.13) \qquad \begin{cases} A_j \subset W_j, \ \overline{W_j} \subset O\backslash F, \ \text{diam}\, \overline{W_j} < \epsilon \text{ for } 1 \leq j \leq m. \\ \{\overline{W}_1, \ldots, \overline{W}_m\} \text{ is a succession of closed sets.} \end{cases}$$

(b) *There exists $h \in H$ such that*

$$(5.14) \qquad \begin{cases} h = 1_X \quad \text{on} \quad X\backslash(\cup_{j=1}^m W_j) \\ h(x) \in O'. \end{cases}$$

PROOF. By shrinking ϵ if necessary we can assume that $V_{2\epsilon}(B) \subset O$. Now apply Lemmas 4.2 and 4.5 to choose a positive $\delta < \epsilon$ such that any pair of points in $H(x)$ at most δ apart can be joined by a sponge of diameter less than ϵ, and the sponge can be chosen to avoid a finite set.

Because B is connected and $x, x' \in B$ they can be joined by δ-chains for 1_B. That is, we can choose a sequence $\{y_0, \ldots, y_n\}$ in B with $y_0 = x$, $y_n = x'$ and $d(y_{i-1}, y_i) < \delta$ for $1 \leq i \leq n$. Since $B \subset \overline{H(X)}$ we can perturb slightly to get a sequence $\{x_0, \ldots, x_n\}$ in $H(x) \cap V_\epsilon(B)$ with $x_0 = x, x_n = x'$ and $d(x_{i-1}, x_i) < \delta$. By Proposition 4.8 (b) $H(x)$ has no isolated points and so we can make sure that $\{x_0, \ldots, x_n\}$ is disjoint from F. Choose $\widetilde{A}_i$ a sponge containing x_{i-1} and x_i for $1 \leq i \leq n$, with $\widetilde{A}_i \cap F = \emptyset$ and diam $\widetilde{A}_i < \epsilon$.

Refine the sequence to get a succession by defining $i(1) = 1$ and, inductively,

$$i(j+1) = \max\{i : \widetilde{A}_{i(j)} \cap \widetilde{A}_i \neq \emptyset\}. \tag{5.15}$$

Observe that if $i(j) < n$ then $i(j+1) \geq i(j)+1$. So the process continues until for some m, $i(m) = n$ at which point it stabilizes. Write A_j for $\widetilde{A}_{i(j)}$ with $j = 1, \ldots, m$. Note that $x \in A_1 = \widetilde{A}_1$ and $x' \in A_m = \widetilde{A}_n$. $\{A_1, \ldots, A_m\}$ is clearly a succession.

Now inductively choose an open set W_j containing A_j, with $\overline{W}_j \cap F = \emptyset$, diam $\overline{W}_j < \epsilon$ and such that:

$$\begin{cases} \overline{W}_j \cap \overline{W}_i = \emptyset \text{ for } i < j-1 \\ \overline{W}_j \cap A_i = \emptyset \text{ for } i > j+1. \end{cases} \tag{5.16}$$

It follows that $\{\overline{W}_1, \ldots, \overline{W}_m\}$ is a succession of sets disjoint from F. Since $V_{2\epsilon}(B) \subset O$ and each A_j meets $V_\epsilon(B)$ we can choose $W_j \subset O$ for all j.

For part (b) choose G_j a crushing neighborhood for A_j in W_j $(j = 1, \ldots, m)$. Now for $1 \leq j < m$ let $h_j \in H$ with $h = 1_X$ on $X \backslash W_j$ and $h_j(G_j) \subset G_j \cap G_{j+1}$. Let $h_m \in H$ with $h_m = 1_X$ on $X \backslash W_m$ and $h_m(G_m) \subset O'$. The product $h = h_m \ldots h_1 \in H$ maps x into O' as required. □

The problem arises because we want h to be close to 1 in H. Even if each h_j is close to 1 the product might not be if we have no control over the number m of factors. On the other hand, suppose that O has diameter less than ϵ. Note that $\{h \in H(X) : h = 1_X \text{ on } X \backslash O\}$ is a closed subgroup of $H(X)$ and with respect to the metric (3.8) the subgroup is entirely contained in $V_\epsilon(1_X)$. Since each h_j lies in this subgroup, this product h does as well and so $h \in V_\epsilon(1_X)$ in $H(X)$.

PROPOSITION 5.4. *Let H be a closed subgroup of the automorphism group $H(X)$. Assume that the restriction to H of the evaluation action on X is strictly spongy. Assume that $x \in X$ and the H-iso-class closure $\overline{H(x)}$ has only finitely many components. Let $f \in H$ and $\epsilon > 0$. Assume that $x \in \omega(x, f)$ and that O_0 is an open subset of X with $x \in O_0$. There exists $h \in X$, $U \subset X$ and a positive integer k such that:*

(1) *$h = 1_X$ on $X \backslash O_0$ and $d(h, 1_X) < \epsilon$.*
(2) *U is an inward set for $g = hf$ with an (ϵ, k)-periodic decomposition $\{U_0, \ldots, U_{k-1}\}$ such that $x \in U_0 \subset O_0$. Furthermore, there is a succession $\{Q_1, \ldots, Q_5\}$ such that U_0 is a topological horseshoe for g^k with respect to $\{Q_1, \ldots, Q_5\}$ and $\{Q_1 \cap \overline{H(x)}, \ldots, Q_5 \cap \overline{H(x)}\}$ is a succession with respect to which the restriction of g^k to $\overline{H(x)}$ is a topological horseshoe.*

Moreover, if x is a periodic point for f, the integer k can be chosen to be the period of x.

PROOF. Let X_0 be the closed invariant subset $\overline{H(x)}$. By Lemma 4.7(b) any sponge which meets X_0 is contained in X_0. By Proposition 4.8(a) the restriction of the H action to X_0 is strictly spongy and by Proposition 4.8 (c) applied to this action X_0 is locally connected and every sponge in X_0 is connected.

By shrinking O_0 if necessary we can assume that

$$\text{diam } O_0 < \epsilon. \tag{5.17}$$

Let N be a neighborhood of x such that any two points of $N \cap H(x)$ are contained in a perfect sponge in O_0. Because $x \in \omega(x, f)$ there exists a positive integer j such that $f^j(x) \in N$. Choose a perfect sponge K_0 in O_0 which contains x and $f^j(x)$ where j is the first return time of x to N, i.e. the smallest positive integer j such that $f^j(x) \in N$. Then let k be the smallest positive integer such that $f^k(x) \in K_0$. Thus, we have constructed a perfect sponge K_0 such that

$$(5.18) \qquad \begin{cases} \{x, f^k(x)\} \subset K_0 \subset O_0, \\ f^i(x) \notin K_0 \quad \text{for} \quad 0 < i < k. \end{cases}$$

Notice that if x is a periodic point with period k we can shrink O_0 so that $f^i(x) \notin O_0$ for $0 < i < k$. Then any sponge K_0 in O_0 which contains $x = f^k(x)$ satisfies (5.18).

Choose a positive $\gamma < \epsilon/2$ such that

$$(5.19) \qquad \begin{cases} V_\gamma(K_0) \subset O_0, \\ V_\gamma(K_0) \cap V_\gamma(\{f(x), \ldots, f^{k-1}(x)\}) = \emptyset. \end{cases}$$

Let O_1 be an open crushing neighborhood of K_0 in $V_\gamma(K_0)$ and choose O_2 open and containing x so that

$$(5.20) \qquad \begin{cases} O_2 \cup f^k(O_2) \subset O_1 \\ f^i(O_2) \subset V_\gamma(f^i(x)) \quad \text{for} \quad 0 < i < k. \end{cases}$$

By Proposition 4.8, x is not an isolated point of $H(x)$. So we can choose a point x' in $H(x)$ distinct from x but close enough that the pair is contained in a sponge K_1 in O_2. That is,

$$(5.21) \qquad \begin{cases} \{x, x'\} \subset K_1 \subset O_2 \\ x' \in H(x) \cap (O_2 \backslash \{x\}). \end{cases}$$

Because the sponge K_1 is a connected subset of X, we can apply Lemma 5.3 to obtain sponges A_j such that:

$$(5.22) \qquad \begin{cases} \{A_1, \ldots, A_m\} \text{ is a succession of sponges } x \in A_1, x' \in A_m \\ A_j \cap H(x) \neq \emptyset \quad \text{for} \quad 1 \leq j \leq m \end{cases}$$

and open sets W_j such that

$$(5.23) \qquad \begin{cases} A_j \subset W_j \quad \text{for} \quad 1 \leq j \leq m \\ W \underset{\text{def}}{=} \bigcup_{j=1}^m W_j \subset\subset O_2 \\ \{\overline{W_1}, \ldots, \overline{W_m}\} \text{ is a succession of closed subsets.} \end{cases}$$

Furthermore, we can assume that the diameters satisfy

$$\text{diam } \overline{W_j} < d(x, x')/5. \tag{5.24}$$

This clearly forces

$$m \geq 5. \tag{5.25}$$

In particular, $\overline{W_3}$ is defined and disjoint from $\overline{W_1}$ and $\overline{W_m}$. By shrinking W_3 if necessary we can assume:

$$W_3 \text{ is a crushing neighborhood for } A_3 \text{ in } O_2 \backslash (\overline{W_1} \cup \overline{W_m}). \tag{5.26}$$

Define G_1 and G_m:

$$G_j \text{ is a crushing neighborhood for } A_j \text{ in } W_j \text{ for } j = 1, m. \tag{5.27}$$

Since each sponge A_j meets $H(x)$ by (5.22), it follows that each A_j is connected and so

$$A \underset{\text{def}}{=} \bigcup_{j=1}^{m} A_j \text{ is connected.} \tag{5.28}$$

By Proposition 4.8 (c), $\overline{H(x)}$ is locally connected. Hence, the A_1 component of $G_1 \cap \overline{H(x)}$ is open relative to $\overline{H(x)}$. So we can shrink G_1 to assume

$$G_1 \cap \overline{H(x)} \text{ is connected.} \tag{5.29}$$

Now choose a point x'':

$$x'' \in A_3 \cap H(x). \tag{5.30}$$

Since $x \in G_1 \cap K_0$ we can define a crushing map $h_0 \in H$:

$$\begin{cases} h_0 = 1_X \quad \text{on} \quad X \backslash V_\gamma(K_0) \\ h_0(O_1) \subset\subset G_1. \end{cases} \tag{5.31}$$

With $g_0 = h_0 f$, (5.19) and (5.20) imply that

$$\begin{cases} g_0^i = f^i \text{ on } O_2 \text{ for } 0 \leq i < k \\ g_0^k = h_0 f^k \text{ on } O_2. \end{cases} \tag{5.32}$$

By (5.20) $f^k(O_2) \subset O_1$ and so

$$g_0^k(O_2) \subset\subset G_1. \tag{5.33}$$

By (5.23), g_0^k maps $\overline{W}$, the union of the succession of neighborhoods into $G_1 \subset W_1$. In particular, $\overline{W}$ is inward for g_0^k. Moreover, it is crushed by g_0^k into the x end of the succession. We perturb to get a horseshoe by dragging the image of the middle point x'' back to the x' end of the succession. The dragging is accomplished by using Lemma 5.3 again. See Figure 5.1.

This time we use the connected set $B = (\overline{G_1 \cap H(x)}) \cup A$ contained in the open set W. The two points we will consider are $g_0^k(x'')$ and x' , with the latter contained in the open set G_m. We will avoid the finite set $F = \{g_0^k(x), g_0^k(x')\}$. By the lemma there exists an open set O_3 and a homeomorphism $h_1 \in H$ such that

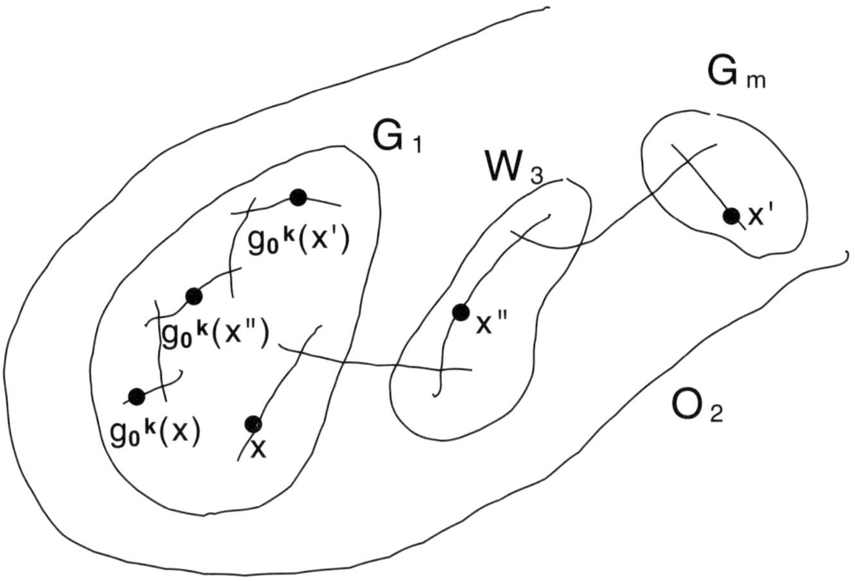

FIGURE 5.1

$$(5.34)\qquad \begin{cases} O_3 \subset\subset W\backslash\{g_0^k(x), g_0^k(x')\} \\ h_1 = 1_X \quad \text{on} \quad X\backslash O_3 \\ h_1(g_0^k(x'')) \in G_m. \end{cases}$$

There will be a final bit of crushing. Define open sets O_4, O_4', O_4'' so that:

$$(5.35)\qquad \begin{cases} x \in O_4 \subset\subset G_1 \cap g_0^{-k}(G_1\backslash\overline{O}_3) \\ x' \in O_4' \subset\subset G_m \cap g_0^{-k}(G_1\backslash\overline{O_3}) \\ x'' \in O_4'' \subset\subset W_3 \cap (h_1 g_0^k)^{-1}(G_m). \end{cases}$$

Define crushing maps $h_2, h_2' \in H$ for G_1 and G_m so that:

$$(5.36)\qquad \begin{cases} h_2 = 1_X \text{ on } X\backslash W_1, \; h_2' = 1_X \text{ on } X\backslash W_m \\ h_2(G_1) \subset\subset O_4, \; h_2'(G_m) \subset\subset O_4'. \end{cases}$$

The final crushing map $h_3 \in H$ for W_3 satisfies:

$$(5.37)\qquad \begin{cases} h_3 = 1_X \text{ on } (X\backslash O_2) \cup \overline{W}_1 \cup \overline{W}_m \\ h_3(W_3) \subset\subset O_4''. \end{cases}$$

Let $h = h_3 h_2' h_2 h_1 h_0$ and $g = hf$. Observe that as in (5.32) we have:

$$(5.38)\qquad \begin{cases} g^i = f^i \text{ on } O_2 \text{ for } 1 \le i < k \\ g^k = hf^k \text{ on } O_2. \end{cases}$$

Now we define the pieces for our horseshoe:

$$(5.39)\qquad \begin{cases} Q_1 = \overline{O_4},\ Q_3 = \overline{O_4''},\ Q_5 = \overline{O_4'} \\ Q_2 = h_3(\overline{W}_1 \cup \overline{W}_2),\ Q_4 = h_3(\bigcup_{j=4}^m \overline{W}_j). \end{cases}$$

Contained as they are in W_1, W_3 and W_m, the sets Q_1, Q_3 and Q_5 are pairwise disjoint. Since the $\overline{W}_j$'s form a succession and h_3 is a homeomorphism Q_2 and Q_4 are disjoint. For the same reason $Q_1 \subset W_1 = h_3(W_1)$ is disjoint from Q_4 and similarly Q_5 is disjoint from Q_2. On the other hand, $h_3(W_3) \subset Q_3$ implies:

$$(5.40)\qquad h_3(A) \subset h_3(W) \subset h_3(W) \cup Q_3 = Q$$

where $Q \underset{\text{def}}{=} \cup_{i=1}^5 Q_i$. Since $h_3(A)$ is connected and meets Q_1 and Q_5 it follows that Q_1 and Q_5 cannot be separated in Q. Hence, $\{Q_1, \ldots, Q_5\}$ is a connected succession.

Observe next that by (5.34) and (5.36) h_1, h_2 and h_2' equal 1_X on $X \backslash W$. Hence each preserves W. So by (5.33) we have

$$(5.41)\qquad \begin{cases} g^k(O_2) = h_3(h_2' h_2 h_1) g_0^k(O_2) \subset\subset \\ h_3(h_2' h_2 h_1)(W_1) \subset h_3(W) \subset Q. \end{cases}$$

By (5.37) h_3 preserves O_2 and so

$$(5.42)\qquad Q \subset h_3(W) \cup W_3 \subset O_2.$$

So we have proved that Q is inward for g^k.

Now we consider the pieces Q_1, Q_3 and Q_5. By (5.35), g_0^k maps $Q_1 = \overline{O_4}$ into $G_1 \backslash \overline{O_3}$. Then h_1 ignores the image since it is the identity on $X \backslash O_3$. h_2 then crushes it back into Q_1. The resulting image is ignored by h_2' and h_3. Similarly, g_0^k maps Q_5 to $G_1 \backslash \overline{O_3}$. The image is ignored by h_1 crushed into Q_1 by h_2 and then ignored by h_2' and h_3. So we have:

$$(5.43)\qquad g^k(Q_1 \cup Q_5) \subset\subset Q_1.$$

By (5.35) $h_1 g_0^k$ maps Q_3 into G_m. The image is ignored by h_2 and h_2' crushes it into Q_5. The image is then ignored by h_3. Hence,

$$(5.44)\qquad g^k(Q_3) \subset\subset Q_5.$$

Thus, Q is a topological horseshoe for g^k. Furthermore, since $A \subset X_0$, $g^k|(Q \cap X_0)$ is a horseshoe with succession $\{Q_1 \cap X_0, \ldots, Q_5 \cap X_0\}$.

Since $Q \subset O_2$ by (5.42), it follows from (5.20) that $f^i(Q) = g^i(Q)$ is contained in $Q_\gamma(f^i(x))$ for $0 \le i < k$. Let $U_0 = U_k = Q$. For $i = k-1, \ldots, 1$ we proceed by downward induction to define closed sets U_i so that

$$(5.45)\qquad g^i(Q) \subset\subset U_i \subset\subset V_\gamma(f^i(x)) \cap g^{-1}(U_{i+1}).$$

$\{U_i\}$ is a k-periodic decomposition for the inward set $U = \bigcup_{j=1}^{k-1} U_i$ for g. Since $2\gamma < \epsilon$ it is an (ϵ, k)-periodic decomposition. By construction, $U_0 = Q$. □

We conclude by observing that by using longer successions one can define other sorts of horseshoes with more back and forth twists. For each such pattern one can prove a version of Proposition 5.2 leading to chain components of f^k which map onto an n shift for arbitrarily large n. With the same hypotheses as in Proposition 5.4 one can use an analogue of the proof to obtain any such generalized horseshoe pattern. In particular, the topological entropy of such a homeomorphism is then at least as big as $\ln(n)/k$, because the entropy of f^k is k times the entropy of f, the entropy of f maximizes the entropy of the restriction of f to any closed subset, the entropy of the extension f^k is at least as great as the entropy of the n-shift, which is well known to be $\log(n)$ (see Denker, Grillenberger and Sigmund (1976) for a definition of topological entropy and proofs of these well-known facts). Note that n can be increased indefinitely without changing k, so we have outlined a way of generating homeomorphisms with arbitrarily large topological entropy.

CHAPTER 6

Generic Homeomorphisms

Overview: Here we harvest the applications of the arguments sown in the previous sections. We describe several classes of homeomorphisms and show that, under appropriate hypotheses, these are residual. We show that if f is an element of these residual sets, then it has many of the properties described in the Introduction.

Throughout this section H is a Polish group with $e : H \times X \to X$ an action on the space X. We will continue to identify an element $h \in H$ with the homeomorphism $e^{\#}(h) \in H(X)$, where $e^{\#} : H \to H(X)$ is the continuous homomorphism associated with the action. So, for example, if $G \subset H(X)$ we will write $\{h \in G\}$ to mean $(e^{\#})^{-1}(G) \subset H$. $\mathcal{U}$ is assumed to be an H-invariant basis for X, i.e., if $h \in H$ and $O \in \mathcal{U}$ then $h(O) \in \mathcal{U}$.

6.1. The classes $\mathcal{H}_1$ and $\mathcal{H}_{1,s}$

We begin by defining two families of subsets of H. Recall the definition of the prolongational nonwandering set $PNW_H(f)$ from (3.27).

For closed subsets K, K_1 of X, $\epsilon > 0$ and m a positive integer we define subsets $G^1(K, K_1, \epsilon, m, \mathcal{U})$ and $G^2(K, \epsilon, m, \mathcal{U})$ of H:

(6.1) $\qquad f \in G^1(K, K_1, \epsilon, m, \mathcal{U}) \Leftrightarrow$ either 6.1.1 or 6.1.2 is satisfied:

(6.1.2) : $K_1 \cap [\cap_{i=0}^{\infty} f^{-i}(K)] = \emptyset$,
(6.1.2) : $V_\epsilon(K)$ contains U, an (ϵ, k)-periodic inward set of $\mathcal{U}$ type for f with $k \geq m$, such that $V_\epsilon(K_1)$ meets the proper basin of the associated attractor $\omega(U, f)$.

Similarly

(6.2) $\qquad f \in G^2(K, \epsilon, m, \mathcal{U}) \Leftrightarrow$ either of the conditions 6.2.1, 6.2.2 is met:

(6.2.1) : $K \cap PNW_H(f) = \emptyset$,
(6.2.2) : $V_\epsilon(K)$ contains a piece of a $\mathcal{U}$ type decomposition for an (ϵ, k)-periodic inward set for f with $k \geq m$.

LEMMA 6.1. *Each $G^1(K, K_1, \epsilon, m, \mathcal{U})$ and $G^2(K, \epsilon, m, \mathcal{U})$ is an open subset of H. If $\widetilde{K}, \widetilde{K}_1$ are closed subsets of X such that $K \subset \widetilde{K} \subset \overline{V_\epsilon(K)}$ and $K_1 \subset \widetilde{K}_1 \subset \overline{V_\epsilon(K_1)}$ then*

$$(6.3) \qquad \begin{cases} G^1(\widetilde{K}, \widetilde{K}_1, \epsilon, m, \mathcal{U}) \subset G^1(K, K_1, 2\epsilon, m, \mathcal{U}) \\ \quad G^2(\widetilde{K}, \epsilon, m, \mathcal{U}) \subset G^2(K, 2\epsilon, m, \mathcal{U}) \end{cases}$$

If the action of H on X is spongy then each of the sets

$$G^1(K, K_1, \epsilon, 1, \mathcal{U}) \text{ and } G^2(K, \epsilon, 1, \mathcal{U})$$

is dense in H. If the action of H on X is strictly spongy then each of the sets $G^1(K, K_1, \epsilon, m, \mathcal{U})$ and $G^2(K, \epsilon, m, \mathcal{U})$ is dense in H.

PROOF. We first observe that each of the four defining conditions is open.

If $K_1 \cap [\cap_{i=0}^{\infty} f^{-i}(K)] = \emptyset$ then by compactness $K_1 \cap [\cap_{i=0}^{N} f^{-i}(K)] = \emptyset$ for sufficiently large N. In particular, there is a positive constant δ such that these two compact sets are more than δ apart. By continuity, if f_1 is sufficiently close to f, then each of the maps f_1^i will be within δ of f^i, $1 \leq i \leq N$, so for all such f_1, $K_1 \cap [\cap_{i=0}^{\infty} f_1^{-i}(K)] = \emptyset$.

If $K \cap PNW_H(f) = \emptyset$ then for some $\epsilon > 0$, $K \cap \overline{[\cup_{d(f,f_1)<\epsilon} \Omega(f_1)]} = \emptyset$. So if $d(f, f_1) < \epsilon$, $K \cap PNW_H(f_1) = \emptyset$.

If $U \subset V_\epsilon(K)$ is an (ϵ, k)-periodic inward set for f then it is an (ϵ, k)-periodic inward set with the same decomposition, for f_1 close enough to f. If $x \in V_\epsilon(K_1)$ is in the proper basin, then for large enough N, $f^N(x) \in U^\circ$ and $x \notin f^N(U)$. For any fixed N these conditions remain true when f is replaced by a close enough f_1. Then x is in the proper f_1 basin.

Similarly, if U is an inward set for f with $\{U_0, \ldots, U_{k-1}\}$ a (ϵ, k)-periodic decomposition of U such that $U_0 \subset V_\epsilon(K)$, then for f_1 close enough to f, U is inward for f_1 with the same decomposition.

An (ϵ, k)-periodic set is $(2\epsilon, k)$-periodic and $V_\epsilon(\widetilde{K}) \subset V_{2\epsilon}(K)$. So (6.3) follows from the definitions (6.1) and (6.2).

Now assume that the action is spongy. Let $f \in H$ and $\epsilon_1 > 0$.

Suppose $x \in K_1 \cap [\cap_{i=0}^{\infty} f^{-i}(K)]$, i.e. $x \in K_1$ and the orbit of x remains in the closed set K. The open set $O = V_\epsilon(K)$ contains $\omega(x, f)$ and we can choose $O_0 \subset O$ open, such that O_0 meets $\omega(x, f)$. Now apply the trapping version of the crushing argument, Proposition 4.10, to get $g \in H$ such that $g \in G^1(K, K_1, \epsilon, 1, \mathcal{U})$ and $d(f, g) < \epsilon_1$.

Suppose that $x \in K \cap PNW_H(f)$. There exists $f_1 \in H$ with $d(f, f_1) < \epsilon/2$ such that $V_\epsilon(x) \cap \Omega(f_1) \neq \emptyset$. Choose $O_0 \subset V_\epsilon(x)$ open such that O_0 meets $\Omega(f_1)$. Apply the closing argument, Proposition 4.9, to get $g \in H$ such that $g \in G^2(K, \epsilon, 1, \mathcal{U})$ and $d(f_1, g) < \epsilon_1/2$ and so $d(f, g) < \epsilon_1$.

In the strictly spongy case we complete the proof using induction on m. By the induction hypothesis we can choose $f_1 \in H$ with $d(f, f_1) < \epsilon_1/2$ and such that

$$f_1 \in G^1(K, K_1, \epsilon, m, \mathcal{U}) \cap G^2(K, \epsilon, m, \mathcal{U}). \tag{6.4}$$

We show that by a perturbation we can increase m in either G^1 or G^2.

Suppose that $V_\epsilon(K)$ contains an (ϵ, k_1)-periodic set V inward for f_1 with $k_1 \geq m$ and $x \in V_\epsilon(K_1)$ meets the basin of the attractor $A_1 = \omega(V, f_1)$, i.e. $\omega(x, f) \subset A_1$. By the period multiplier version of the crushing argument, Proposition 4.11, we can construct g_1 in H with $d(g_1, f_1) < \epsilon_1/2$ and U an (ϵ, k)-periodic set of $\mathcal{U}$ type inward for g_1 with $U \subset V$, $k \geq 2k_1$ and such that x is in the proper basin, of the attractor $A = \omega(U, g_1)$. Thus, $g_1 \in G^1(K, K_1, \epsilon, 2m, \mathcal{U})$.

Similarly, if $V_\epsilon(K)$ contains the piece V_0 of the (ϵ, k_1)-periodic decomposition of V then we can again apply Proposition 4.11 with x a recurrent point in $A_1 \cap V_0$.

We get an inward set U constructed for $g_2 \in H$ with $d(g_2, f_1) < \epsilon_1/2$ which satisfies $U_0 \subset V_0$ and so $g_2 \in G^2(K, \epsilon, 2m, \mathcal{U})$. □

Now we are ready to define the classes of homeomorphisms which we will examine. They are the classes $\mathcal{H}_1[\mathcal{U}]$ and $\mathcal{H}_{1,s}[\mathcal{U}]$ given by the following.

(6.5) $\mathcal{H}_1[\mathcal{U}]$ is the set of all $f \in H$ satisfying 6.5.1:

(6.5.1): For all nonempty closed subsets K, K_1 of X and all $\epsilon > 0$, both f and f^{-1} are in $G^1(K, K_1, \epsilon, 1, \mathcal{U}) \cap G^2(K, \epsilon, 1, \mathcal{U})$.

(6.6) $\mathcal{H}_{1,s}[\mathcal{U}]$ is the set of all $f \in H$ satisfying 6.6.1:

(6.6.1) : For all nonempty closed subsets K, K_1 of X, all $\epsilon > 0$ and all integers $m > 0$, both f, and f^{-1} are in the intersection $G^1(K, K_1, \epsilon, m, \mathcal{U}) \cap G^2(K, \epsilon, m, \mathcal{U})$.

Clearly, if $\mathcal{U}_1$ is another H-invariant basis with $\mathcal{U} \subset \mathcal{U}_1$ we have

$$\begin{array}{ccc} \mathcal{H}_{1,s}[\mathcal{U}] & \subset & \mathcal{H}_1[\mathcal{U}] \\ \cap & & \cap \\ \mathcal{H}_{1,s}[\mathcal{U}_1] & \subset & \mathcal{H}_1[\mathcal{U}_1] \end{array} \tag{6.7}$$

THEOREM 6.2. *For the action of the Polish group H on the space X, the sets $\mathcal{H}_1[\mathcal{U}]$ and $\mathcal{H}_{1,s}[\mathcal{U}]$ are G_δ subsets of H. Furthermore, they are conjugacy invariant subsets, i.e.*

$$f \in \mathcal{H}_1[\mathcal{U}] \text{ and } h \in H \Rightarrow hfh^{-1} \in \mathcal{H}_1[\mathcal{U}], \tag{6.8}$$

with the analogous implication for $\mathcal{H}_{1,s}[\mathcal{U}]$.

If the action of H on X is spongy then $\mathcal{H}_1[\mathcal{U}]$ is dense in H. If the action is strictly spongy then $\mathcal{H}_{1,s}[\mathcal{U}]$ is dense.

PROOF. Suppose in the definitions (6.5) and (6.6) we were to further restrict to ϵ rational and to K, K_1 chosen among closures of finite unions of elements of some countable base for X. Because the number of conditions imposed is then countable the resulting subsets of H would be G_δ by Lemma 6.1. On the other hand, (6.3) shows that allowing more general K, K_1 and ϵ does not affect the resulting intersection.

Hence, $\mathcal{H}_1[\mathcal{U}]$ and $\mathcal{H}_{1,s}[\mathcal{U}]$ are G_δ subsets of H. Because H is Baire, Lemma 6.1 implies that $\mathcal{H}_1[\mathcal{U}]$ (resp. $\mathcal{H}_{1,s}[\mathcal{U}]$) is dense in H when the action is spongy (resp. strictly spongy).

Now fix $h \in H$ and $\epsilon > 0$. Choose $\delta > 0$ an ϵ modulus of uniform continuity for h and h^{-1}. If $\{U_i\}$ is a (δ, k)-periodic decomposition for an f inward set U, then $\{h(U_i)\}$ is an (ϵ, k)-periodic decomposition for the hfh^{-1} inward set $h(U)$. Furthermore if $A \subset V_\delta(B)$ then $h(A) \subset V_\epsilon(h(B))$. It follows that

$$f \in G^1(h^{-1}(K), h^{-1}(K_1), \delta, m, \mathcal{U}) \Rightarrow hfh^{-1} \in G^1(K, K_1, \epsilon, m, \mathcal{U}) \tag{6.9}$$

with a similar implication for G^2. Implication (6.8) easily follows. □

When $\mathcal{U}$ consists of the basis of all open subsets we will omit it as a label writing simply $\mathcal{H}_1$ and $\mathcal{H}_{1,s}$.

THEOREM 6.3. *For the action of* H *on* X, *let* $f \in \mathcal{H}_1$ *as defined by* (6.5) (*with* $\mathcal{U}$ *the basis of all open subsets of* X).

(a) *Let* K *be a closed subset of* X *with* $x \in K$ *and let* $\epsilon > 0$. *If* $f(K) \subset K$ *then* $V_\epsilon(K)$ *contains an attractor* A *such that the proper basin* $W^s_f(A)\backslash A$ *meets* $V_\epsilon(x)$. *If* $f^{-1}(K) \subset K$ *then* $V_\epsilon(K)$ *contains a repellor* R *such that the proper basin* $W^u_f(R)\backslash R$ *meets* $V_\epsilon(x)$.

(b) *If* A *is a nonempty attractor for* f *and* R *is a nonempty repellor for* f *then* $A^\circ \neq \emptyset$ *and* A *contains repellors, and* $R^\circ \neq \emptyset$ *and* R *contains attractors. In fact,*

$$(6.10) \qquad \begin{cases} A = \overline{\cup\{W^u_f(R_1)\backslash R_1 : R_1 \subset A \text{ and } R_1 \text{ is a repellor}\}} \\ R = \overline{\cup\{W^s_f(A_1)\backslash A_1 : A_1 \subset R \text{ and } A_1 \text{ is an attractor}\}}. \end{cases}$$

(c) *Each attractor and each repellor, when nonempty, contain a countable infinity of distinct attractors and repellors and uncountably many chain components.*

PROOF. (a): Obvious because $f, f^{-1} \in G^1(K, x, \epsilon, 1, \mathcal{U})$.

(b): Let A be an attractor and $x \in A$. Choose U inward for f so that $\omega(U, f) = A$ and choose $\epsilon > 0$ small enough that $V_\epsilon(A) \subset U$. Because an attractor for f is f-invariant and so is f^{-1}-invariant we can apply (a) with $x \in K = A$ to find a repellor $R_1 \subset V_\epsilon(A)$ whose proper basin meets $V_\epsilon(x)$.

The repellor R_1 is f-invariant and is contained in U. Hence, $R_1 \subset \omega(U, f) = A$. Now if $z \in X\backslash A$, the basin of the repellor dual to A, then $\alpha(z, f)$ $(= \omega(z, f^{-1}))$ is disjoint from A and *a fortiori* from R_1. Contrapositively, the proper basin $W^u_f(R_1)\backslash R_1$ is contained in A. Furthermore, this open subset of X meets $V_\epsilon(x)$. As x in A and $\epsilon > 0$ were arbitrary we see that the family of open sets $\{W^u_f(R_1)\backslash R_1 \subset A$ and R_1 is a repellor$\}$ are contained in A, and hence in A°. Furthermore, their union is dense in A, proving the first equation in (6.10).

The rest of (b) consists of the application to f^{-1} of what we have already proved.

(c) Suppose that A is a nonempty attractor. By (6.10) there exists a repellor $R_1 \subset A$ with $W^u_f(R_1)\backslash R_1 \neq \emptyset$. Let A_1 be the attractor dual to R_1. Equation (6.10) also implies that $W^u_f(R_1) \subset A$. Thus, if $x \in W^u_f(R_1)\backslash R_1$ then $\omega(x, f) \subset A \cap A_1$ and so $\widetilde{A}_1 \equiv A \cap A_1$ is a nonempty attractor contained in A. Also, $\alpha(x, f) \subset R_1$ and so R_1 is nonempty. Now choose A_0 a nonempty attractor contained in R_1. We have constructed a disjoint pair of attractors, $A_0, \widetilde{A}_1$, contained in A. Iterating this process we get at the nth stage 2^n disjoint attractors contained in A. This procedure yields countably many distinct attractors in A and uncountably many distinct decreasing sequences of attractors in A, with intersections containing distinct chain components. $\square$

Recall the sets Π_∞ and $\Pi_{\infty,\infty}$ defined in (2.6) and (2.7).

THEOREM 6.4. *For the action of* H *on* X, *let* $f \in \mathcal{H}_1[\mathcal{U}]$ *as defined in* (6.5). *Then the chain recurrent set* $\mathcal{C}(f)$ *is nowhere dense in* X. *Additionally,*

(a) *The nonwandering set* $\Omega(f)$ *is a perfect, nowhere dense subset of* X. *Both* $\Pi_\infty(f, \mathcal{U})$, $\Pi_\infty(f^{-1}, \mathcal{U})$ *and their intersection are* G_δ *sets, dense in* $\Omega(f)$. *In particular, the minimal points are dense in* $\Omega(f)$.

(b) *The stable and unstable sets* $W^s_f(\Pi_\infty(f,\mathcal{U})), W^u_f(\Pi_\infty(f^{-1},\mathcal{U}))$ *and their intersection are* G_δ *sets, dense in* X.

(c) *If* A, R *is an attractor-repellor pair with proper basin* $X\backslash(A\cup R) = W^s_f(A)\backslash A = W^u_f(R)\backslash R$, *then*

$$(X\backslash(A\cup R))\cap W^s_f(\Pi_\infty(f,\mathcal{U}))\cap W^u_f(\Pi_\infty(f^{-1},\mathcal{U}))$$

is a G_δ *set, dense in the open set* $X\backslash(A\cup R)$. *For* x *in this residual subset,* $\omega(x,f)$ *is a terminal chain component with* $\omega(x,f)\subset(\partial A)\cap\Pi_\infty(f,\mathcal{U})$ *and* $\alpha(x,f)$ *is an initial chain component with* $\alpha(x,f)\subset(\partial R)\cap\Pi_\infty(f^{-1},\mathcal{U})$. *Furthermore,* $\alpha(x,f)\neq\omega(x,f)$.

(d) *The quotient space* $\mathcal{B}_f$ *of chain components is perfect and so is a Cantor set. The collection of chain components that are in both* $\Pi_\infty(f,\mathcal{U})$ *and* $\Pi_\infty(f^{-1},\mathcal{U})$ *is a dense* G_δ *in* $\mathcal{B}_f$.

(e) *If* $f\in\mathcal{H}_{1,s}[\mathcal{U}]$ *then* (a) – (d) above remain true when Π_∞ is replaced by $\Pi_{\infty,\infty}$ throughout.

PROOF. $\mathcal{C}(f)$ is nowhere dense: For any attractor-repellor pair (A,R), $\mathcal{C}(f)\subset A\cup R$ and so the proper basin is an open set disjoint from $\mathcal{C}(f)$. For a homeomorphism the entire space X is an attractor and a repellor with dual $\emptyset$. By (6.10) applied to $R = X$, the proper basins form a dense open set disjoint from $\mathcal{C}(f)$.

(b) Given x and a positive integer n, $f\in G^1(X,x,1/n,1,\mathcal{U})$ implies $V_{1/n}(x)$ meets $W^s_f(\Pi_{(n,k)}(f,\mathcal{U}))$ for some positive integer k. So for all n, the set

$$\cup_{k=1}^\infty W^s_f(\Pi_{(n,k)}(f,\mathcal{U}))$$

is dense in X, and it is open by (2.12). Consequently, by (2.13), $W^s_f(\Pi_\infty(f,\mathcal{U}))$ is residual. For $W^u_f(\Pi_\infty(f^{-1},\mathcal{U}))$ apply the result to f^{-1}.

In addition, if $f\in\mathcal{H}_{1,s}[\mathcal{U}]$ then $f\in G^1(X,x,1/n,m,\mathcal{U})$ for all positive integers m implies that k can be chosen arbitrarily large in that case. So $W^s_f(\Pi_{\infty,\infty}(f,\mathcal{U}))$ and $W^u_f(\Pi_{\infty,\infty}(f^{-1},\mathcal{U}))$ are residual when $f\in\mathcal{H}_{1,s}[\mathcal{U}]$.

(c) Since $X\backslash(A\cup R)$ is open, its intersection with the dense G_δ subset of X,

$$W^s_f(\Pi_\infty(f,\mathcal{U}))\cap W^u_f(\Pi_\infty(f^{-1},\mathcal{U})),$$

is a G_δ set dense in $X\backslash(A\cup R)$. For x in the intersection, $\omega(x,f)$ is a terminal chain component in $\Pi_\infty(f)$ by Corollary 2.4. As $x\in W^s_f(A)$, $\omega(x,f)\subset A$. But the orbit of x remains in the invariant set $X\backslash(A\cup R)$ and so $\omega(x,f)$ is in ∂A, the topological boundary of A. Similarly, $\alpha(x,f)$ is an initial chain component in ∂R. These chain components are distinct because A and R are disjoint. Again if $f\in\mathcal{H}_{1,s}[\mathcal{U}]$ we can replace Π_∞ by $\Pi_{\infty,\infty}$ by using the replacement in (b).

(a): Given $x\in\Omega(f)$ and $\epsilon>0$, $f\in G^2(x,\epsilon,1)$ implies there is an attractor A with periodic decomposition $\{A_0,\dots,A_{k-1}\}$ such that $A_0\subset V_\epsilon(x)$. By (6.10) there is a repellor R_1 such that $W^u_f(R_1)\backslash R_1$ is a nonempty subset of A.

Apply (c). Choose

$$x_1\in(W^u_f(R_1)\backslash R_1)\cap W^s_f(\Pi_\infty(f,\mathcal{U}))\cap W^u_f(\Pi_\infty(f^{-1},\mathcal{U})).$$

As above, $\alpha(x_1,f)\subset R_1\cap\Pi_\infty(f^{-1},\mathcal{U})$ and $\omega(x_1,f)\subset\Pi_\infty(f,\mathcal{U})$. Clearly, $\omega(x_1,f)\subset A$ as well. By Lemma 2.1(a) $\omega(x_1,f)\cap A_0$ and $\alpha(x_1,f)\cap A_0$ are nonempty and so $V_\epsilon(x)\cap\Pi_\infty(f,\mathcal{U})$ and $V_\epsilon(x)\cap\Pi_\infty(f^{-1},\mathcal{U})$ are nonempty. $\Pi_\infty(f)$ and $\Pi_\infty(f^{-1})$ are contained in $\Omega(f)$ since by Proposition 2.2 each is a union of minimal sets.

They are G_δ sets by Lemma 2.5. As ϵ was arbitrary they are residual subsets of $\Omega(f)$. Finally, $\omega(x_1, f) \cap \alpha(x_1, f) = \emptyset$ and so at least one of these sets contains points other than x. Hence, $\Omega(f)$ is perfect. It is nowhere dense because $\mathcal{C}(f)$ is. If $f \in \mathcal{H}_{1,s}[\mathcal{U}]$ we can replace Π_∞ by $\Pi_{\infty,\infty}$ as before.

(d): Recall from Chapter 1 that we use a complete $\mathcal{C}(f)$ Lyapunov function $L : X \to [0,1]$ to define a metric on $\mathcal{B}_f$. Given a chain component B and $\epsilon_1 > 0$ choose $x \in \Omega(f) \cap B$ (e.g. a limit point of any orbit in B) and $\epsilon > 0$ such that $|L(x_1) - L(B)| < \epsilon_1$ for all $x_1 \in V_\epsilon(x)$. Apply the proof above of (a) to get a $\Pi_\infty(f^{-1},\mathcal{U})$ chain component $\alpha(x_1, f)$ and a $\Pi_\infty(f,\mathcal{U})$ chain component $\omega(x_1, f)$ with $\alpha(x_1, f) \neq \omega(x_1, f)$ and each meeting $V_\epsilon(x)$. As L is constant on chain components, L is within ϵ_1 of $L(B)$ on $\omega(x_1, f)$ and on $\alpha(x_1, f)$. Thus, the $\Pi_\infty(f,\mathcal{U})$ and $\Pi_\infty(f^{-1},\mathcal{U})$ chain components are dense in $\mathcal{B}_f$. They are G_δ by Lemma 2.5 again. Since $\omega(x_1, f) \neq \alpha(x_1, f)$ it follows that at least one is not B and so $\mathcal{B}_f$ is perfect. As $\mathcal{B}_f$ is always compact and zero-dimensional it is a Cantor set.

We can again replace Π_∞ by $\Pi_{\infty,\infty}$ in this proof when $f \in \mathcal{H}_{1,s}[\mathcal{U}]$. Thus, we have proved (e) along the way. □

It follows that in the strictly spongy case it is the adding machine chain components in $\Pi_{\infty,\infty}$ which are generic rather than the periodic points of $\Pi_\infty \setminus \Pi_{\infty,\infty}$. The question of the density of periodic points in $\Omega(f)$ is a delicate issue. This may account for the on-and-off status of the claim that for a residual set of homeomorphisms on a smooth manifold the periodic points are dense in the nonwandering set. The question was finally resolved in the affirmative in Hurley (1996b); see that paper for a history of the question. At this point we can generalize that result.

COROLLARY 6.5. *Assume that for X the collection $\mathcal{U}_F$, of open sets whose closure satisfy the fixed point property, is a basis for X. If $f \in \mathcal{H}_1[\mathcal{U}_F]$ then the periodic points of f are dense in $\Omega(f)$.*

PROOF. From (2.24) we have $\Pi_\infty(f,\mathcal{U}_F) \subset \overline{\mathrm{Per}(f)}$. The latter set is contained in $\Omega(f)$. By (a) of Theorem 6.4, $\Pi_\infty(f,\mathcal{U}_F)$ is dense in $\Omega(f)$ when $f \in \mathcal{H}_1[\mathcal{U}_F]$. □

6.2. The class $\mathcal{H}_{3,s}$

We can sharpen the results when the action is generalized homogeneous as well as spongy. Recall from (1.24) that $CT(f)$ denotes the set of closed subsets of X which are chain transitive with respect to the restriction of f. $CT(f)$ is a closed subset of the space $C(X)$ of closed subsets of X when the space is equipped with the Hausdorff metric defined in (1.25). For a closed subset K of X, $\epsilon > 0$ and m a positive integer we define $G^3(K, \epsilon, m, \mathcal{U})$, a subset of H:

$$f \in G^3(K, \epsilon, m, \mathcal{U}) \Leftrightarrow \text{condition 6.11.1 is met:} \tag{6.11}$$

(6.11.1) : There exists U an (ϵ, k)-periodic set of $\mathcal{U}$ type, inward for f with $k \geq m$ and the Hausdorff distance $d(K, U) < \epsilon + d(K, A)$ for all $A \in CT(f)$.

LEMMA 6.6. *Each $G^3(K, \epsilon, m, \mathcal{U})$ is an open subset of H. If $\widetilde{K}$ is a closed subset of X such that $K \subset \widetilde{K} \subset \overline{V_\epsilon(K)}$ then*

$$G^3(\widetilde{K}, \epsilon, m, \mathcal{U}) \subset G^3(K, 3\epsilon, m, \mathcal{U}). \tag{6.12}$$

If the action of H on X is spongy and is generalized homogeneous on a dense subset D of X, then each $G^3(K,\epsilon,m,\mathcal{U})$ is dense in H.

PROOF. Regarding CT as a relation from $H(X)$ to $C(X)$ it is usc, i.e. a closed relation, see GTDS Exercise 7.37(d). So if a closed ball about K in $C(X)$ is disjoint from $CT(f)$ then it is disjoint from $CT(g)$ for all g close enough to f. If we define $d(K, CT(f)) \underset{\text{def}}{=} \inf\{d(K,A) : A \in CT(f)\}$ then it follows that for any positive δ

$$d(K, CT(g)) > d(K, CT(f)) - \delta \tag{6.13}$$

provided g is close enough to f. As an (ϵ, k)-periodic inward set for f is also one for g close enough to f, it follows that $G^3(K,\epsilon,m,\mathcal{U})$ is open.

If $K \subset \widetilde{K} \subset \overline{V_\epsilon(K)}$ then $d(K,\widetilde{K}) \leq \epsilon$ from which (6.12) follows easily.

Now assume that the action H is generalized homogeneous on a dense set D of X, as well as spongy. Recall that the definition requires that D is H-invariant.

Suppose that $A \in CT(f)$ with $d(K,A) = d(K,CT(f))$. Given $\epsilon_1 > 0$ we construct $g \in G^3(K,\epsilon,m,\mathcal{U})$ which is ϵ_1 close to f. By shrinking ϵ_1 if necessary we can assume that $d(f,g) < \epsilon_1$ implies (6.13) with $\delta = \epsilon/3$.

Let $\epsilon_2 = \min(\epsilon/3, \epsilon_1/2)$ and choose $\delta_1 > 0$ an ϵ_2 modulus of generalized homogeneity with $\delta_1 < \epsilon_2$. Because A is chain transitive we can choose a periodic $(\delta_1/2)$-chain $\{x_0,\dots,x_k\}$ in A which is ϵ_2 dense in A, i.e. $d(f(x_{i-1}),x_i) \leq \delta_1/2$ for $i = 1,\dots,k$ with $x_k = x_0$ and $d(A,\{x_0,\dots,x_{k-1}\}) < \epsilon_2$. By repeating the chain several times we can assume that $k \geq m$. Because the action is spongy, X has no isolated points. Furthermore, D is dense in X. So in D we can move each x_i to get $\{\widetilde{x}_0,\dots,\widetilde{x}_k\}$ a periodic δ_1-chain with $d(A,\{\widetilde{x}_0,\dots,\widetilde{x}_{k-1}\}) < \epsilon_2$ and so that $\{\widetilde{x}_0,\dots,\widetilde{x}_{k-1}\} = \{\widetilde{x}_1,\dots,\widetilde{x}_k\}$ as well as $\{f(\widetilde{x}_0),\dots,f(\widetilde{x}_{k-1})\}$ are lists of k distinct points in D.

Choose $h_1 \in H$ with $d(h_1,1) < \epsilon_2$ so that $h_1(f(\widetilde{x}_{i-1})) = \widetilde{x}_i$ for $i = 1,\dots,k$. Thus, $\{\widetilde{x}_0,\dots,\widetilde{x}_{k-1}\}$ is a k periodic orbit for $f_1 = h_1 f$.

Choose a positive $\gamma < \epsilon_2$ so that the sets in $\{V_\gamma(\widetilde{x}_i) : 0 \leq i < k\}$ are pairwise disjoint and let

$$O_0 = \bigcap_{i=0}^{k-1} f_1^{-i}(V_\gamma(\widetilde{x}_i)). \tag{6.14}$$

Apply the prototype crushing argument, Proposition 4.3, to obtain $h_2 \in V_{\epsilon_2}(1_X)$ with $h_2 = 1_X$ on $X \backslash O_0$ and an inward set U for $g = h_2 f_1 = h_2 h_1 f$ with an (ϵ_2,k)-periodic decomposition $\{U_i\}$ of $\mathcal{U}$ type such that $U_0 \subset O_0$.

Since $f_1^i = g^i$ on O_0 for $0 \leq i \leq k-1$, we have $U_i \subset V_\gamma(\widetilde{x}_i)$ for $i = 0,\dots,k-1$. Hence, $d(U,\{\widetilde{x}_0,\dots,\widetilde{x}_{k-1}\}) < \gamma$ and so

$$\begin{aligned} d(K,U) &\leq d(K,A) + d(A,\{\widetilde{x}_0,\dots,\widetilde{x}_{k-1}\}) + d(\{\widetilde{x}_0,\dots,\widetilde{x}_{k-1}\},U) \\ &< d(K,CT(g)) + (\epsilon/3) + \epsilon_2 + \gamma \leq d(K,CT(g)) + \epsilon). \end{aligned}$$

On the other hand, $d(g,f) \leq d(h_1,1_X) + d(h_2,1_X) \leq 2\epsilon_2 \leq \epsilon_1$. □

Recall the definition of $\mathcal{P}_m(f)$ and $\mathcal{C}_m(f)$ in (3.26). Define $\mathcal{H}_{3,s}[\mathcal{U}]$ by

$$f \in \mathcal{H}_{3,s}[\mathcal{U}] \Leftrightarrow f \text{ satisfies 6.15.1 and 6.15.2:} \tag{6.15}$$

(6.15.1) : For all nonempty closed subsets K, K_1 of X, all $\epsilon > 0$, and all integers $m > 0$, both f, and f^{-1} are in the intersection $G^1(K, K_1, \epsilon, m, \mathcal{U}) \cap G^3(K, \epsilon, m, \mathcal{U})$.

(6.15.2) : For all $m > 0$, $\mathcal{P}_m(f) = \mathcal{C}_m(f)$.

It is easy to see that the equation $\mathcal{P}_m(f) = \mathcal{C}_m(f)$ implies the analogous result for f^{-1}: $\mathcal{P}_m(f^{-1}) = \mathcal{C}_m(f^{-1})$.

THEOREM 6.7. *Assume that the action of the Polish group H on the space X is strictly spongy and is generalized homogeneous on a dense subset D of X.*

The set $\mathcal{H}_{3,s}[\mathcal{U}]$ is a G_δ subset dense in H with

$$\mathcal{H}_{3,s}[\mathcal{U}] \subset \mathcal{H}_{1,s}[\mathcal{U}]. \tag{6.16}$$

Furthermore, $\mathcal{H}_{3,s}[\mathcal{U}]$ is a conjugacy invariant subset of H.

PROOF. We use (6.12) and (6.3) to reduce to a countable family of closed sets K and K_1, and to rational ϵ. Proceed just as in Theorem 6.2. Then Lemmas 6.1 and 6.6 together with Proposition 3.7 imply that $\mathcal{H}_{3,s}[\mathcal{U}]$ is a G_δ dense in H.

For any homeomorphism h it is easy to check that

$$\begin{cases} \mathcal{P}_m(hfh^{-1}) = (h \times h \ldots \times h)(\mathcal{P}_m(f)) \\ \\ \mathcal{C}_m(hfh^{-1}) = (h \times h \ldots \times h)(\mathcal{C}_m(f)), \end{cases} \tag{6.17}$$

with $(m+1)$ fold products of h on the right. Using this result conjugacy invariance follows just as in Theorem 6.2.

To prove the inclusion (6.16) we will show that if $f \in \mathcal{H}_{3,s}[\mathcal{U}]$ then $f \in G^2(K, \epsilon, m, \mathcal{U})$. If $K \cap PNW_H(f) \neq \emptyset$ then because $PNW_H(f) \subset \mathcal{C}(f)$, by (3.28), it follows that K meets some chain component B. Since $f \in \mathcal{H}_{3,s}[\mathcal{U}]$ it lies in $G^3(B, \epsilon/4, m, \mathcal{U})$. Hence there exists an $(\epsilon/4, k)$-periodic inward set U of $\mathcal{U}$ type with $k \geq m$ and $d(U, B) < \epsilon/2$. So some piece of the decomposition has distance less than $\epsilon/2$ from a point in $K \cap B$. As this piece has diameter at most $\epsilon/4$, it is contained in $V_\epsilon(K)$. Thus, $f \in G^2(K, \epsilon, m, \mathcal{U})$. □

THEOREM 6.8. *For the action of H on X, let $f \in \mathcal{H}_{3,s}[\mathcal{U}]$ as defined in* (6.15).

(a) *For each $x \in X$, $\mathcal{P}(x, f) = \mathcal{C}(x, f)$ ($\mathcal{P}(x, f)$ is defined in* (3.6)). *In particular, any closed f-invariant subset A of X which is stable, is a quasi-attractor. That is, $\mathcal{P}(A, f) = A$ implies $\mathcal{C}(A, f) = A$. Furthermore, the chain recurrent set is the nonwandering set $\Omega(f)$.*

(b) *The collection of chain components that are in both of $\Pi_{\infty,\infty}(f, \mathcal{U})$ and $\Pi_{\infty,\infty}(f^{-1}, \mathcal{U})$ is a dense G_δ in $CT(f)$. In particular, for any chain transitive subset A for f and any $\epsilon > 0$ there is an adding machine chain component in $\Pi_{\infty,\infty}(f, \mathcal{U}) \cap \Pi_{\infty,\infty}(f^{-1}, \mathcal{U})$ whose Hausdorff distance from A is $< \epsilon$.*

PROOF. (a); The definition of $\mathcal{H}_{3,s}[\mathcal{U}]$ includes the requirement that $\mathcal{P}_1(f) = \mathcal{C}_1(f)$, which is the same as requiring that $\mathcal{P}(x, f) = \mathcal{C}(x, f)$ for all x. If A is stable, then $\mathcal{P}(x, f) \subset A$ for all $x \in A$, so it follows that $\mathcal{C}(x, f) \subset A$ for each $x \in A$, implying that A is an intersection of attractors, see (1.20). The final remark in (a) follows because $\Omega(f) = \{x \mid x \in \mathcal{P}(f(x), x)\}$ and $\mathcal{C}(f) = \{x \mid x \in \mathcal{C}(f(x), x)\}$; see (3.24).

(b): The sets are G_δ in $CT(f)$ by Lemma 2.5. If $A \in CT(f)$ and $\epsilon > 0$ then $f \in G^3(A, \epsilon/4, 1, \mathcal{U})$ implies there is an $(\epsilon/4, k)$-periodic inward set U with $d(A, U) < \epsilon/2$. By Theorem 6.4 (b), (e) we can choose a point $x \in W^s_f(\Pi_{\infty,\infty}(f, \mathcal{U})) \cap U^\circ$.

Then $\omega(x,f)$ is a $\Pi_{\infty,\infty}(f)$ chain component contained in U. By Lemma 2.1(a), $d(U,\omega(x,f)) \leq \epsilon/4$ and so $\omega(x,f)$ in $\Pi_{\infty,\infty}(f)$ is ϵ close to A. Thus, the $\Pi_{\infty,\infty}(f,\mathcal{U})$ chain components form a dense set in $CT(f)$. The result for f^{-1} implies the $\Pi_{\infty,\infty}(f^{-1},\mathcal{U})$ chain components are residual in $CT(f)$ as well. □

6.3. Dynamic isolation

Recall that the sets $\Pi_{\infty,\infty}(f,\mathcal{U})$ and $\Pi_{\infty,\infty}(f^{-1},\mathcal{U})$ are composed of adding machine chain components. The chain components in the intersection of these two sets are rather special. Suppose B is a chain component with

$$B \in \Pi_{\infty,\infty}(f,\mathcal{U}) \cap \Pi_{\infty,\infty}(f^{-1},\mathcal{U})$$

Then B is both terminal and initial and so

$$\mathcal{C}(B,f) = B = \mathcal{C}(B,f^{-1}). \tag{6.18}$$

Such a set is *dynamically isolated*: for any point x outside of B there is $\epsilon > 0$ such that no ϵ-chain for either f or f^{-1} can start in B and reach x. In particular, either $\omega(x,f) = B$ or $\alpha(x,f) = B$ implies that $x \in B$. Yet by Theorems 6.4 and 6.8, when f is in $\mathcal{H}_{3,s}[\mathcal{U}]$ this is the residual case in $\mathcal{B}_f$ and $CT(f)$.

6.4. Attractor boundaries are quasi-attractors

There are also special attractor results for $\mathcal{H}_{3,s}[\mathcal{U}]$.

THEOREM 6.9. *For the action of H on X, let $f \in \mathcal{H}_{3,s}[\mathcal{U}]$ as defined in* (6.15).

(a) *If A, R is an attractor-repellor pair for f then the topological boundary ∂A is a quasi-attractor and ∂R is a quasi-repellor. In fact,*

$$\begin{cases} \omega\mathcal{C}(X\backslash(A\cup R),f) = \mathcal{C}(\partial A,f) = \partial A \\ \omega\mathcal{C}(X\backslash(A\cup R),f^{-1}) = \mathcal{C}(\partial R,f^{-1}) = \partial R, \text{ and} \end{cases} \tag{6.19}$$

$$\begin{cases} \mathcal{C}(A^\circ,f^{-1}) = A^\circ = \cup\{W^u_f(R_1) : R_1 \subset A \text{ and } R_1 \text{ is a repellor}\} \\ \mathcal{C}(R^\circ,f) = R^\circ = \cup\{W^s_f(A_1) : A_1 \subset R \text{ and } A_1 \text{ is an attractor}\}. \end{cases} \tag{6.20}$$

(b) *If B_1, B_2 are distinct chain components with $B_1 \subset \mathcal{C}(B_2,f)$ then there exists an attractor-repellor pair A, R with $B_1 \subset A$ and $B_2 \subset R$. For any such pair, $B_1 \subset \partial A$ and $B_2 \subset \partial R$. Conversely, if A, R is an attractor-repellor pair and B_1, B_2 are chain components with $B_1 \subset \partial A$, $B_2 \subset \partial R$ then there exist chain components $\widetilde{B}_1 \subset \partial A$, $\widetilde{B}_2 \subset \partial R$ such that $B_1 \subset \mathcal{C}(\widetilde{B}_2,f)$ and $\widetilde{B}_1 \subset \mathcal{C}(B_2,f)$. Thus, a chain component is initial iff it is not in the boundary of any attractor. A chain component is terminal iff it is not in the boundary of any repellor.*

PROOF. We begin by defining an *annular set* for an attractor-repellor pair A, R. Let U be an inward set with $A = \omega(U,f)$, so that $U \cap R = \emptyset$. As f is a homeomorphism $f(U)^\circ = f(U^\circ) \supset A$. Define $Q = U\backslash f(U^\circ)$. So Q is a closed set contained in the open set $X\backslash(A\cup R)$, the proper basin of the pair. If $x \in X\backslash(A\cup R)$ then the bi-infinite orbit sequence $\{f^i(x) : i \in \mathbf{Z}\}$ is contained in $X\backslash(A\cup R)$ with $f^i(x) \to A$ as $i \to \infty$ and $f^i(x) \to R$ as $i \to -\infty$. Hence, $f^i(x) \in U$ for i large enough and $f^i(x) \notin U$ for i small enough. If i_0 is the smallest integer such that $f^{i_0}(x) \in U$ then $f^{i_0}(x) \notin f(U^\circ)$ and so $f^{i_0}(x) \in Q$. That is, every bi-infinite orbit in $X\backslash(A \cup R)$ meets Q (sets with this property are sometimes called *fundamental domains*).

(a): Next we show that $x \in X\backslash(A \cup R)$ implies $\omega\mathcal{C}(x, f) \subset \partial A$ and $\omega\mathcal{C}(x, f^{-1}) \subset \partial R$. In general, $\omega\mathcal{C}(x, f) \subset A$ and $\omega\mathcal{C}(x, f^{-1}) \subset R$ by (1.18). Such an x is not chain recurrent, so

$$\omega\mathcal{C}(x, f) = \mathcal{C}(x, f)\backslash\mathcal{O}(x, f) = \mathcal{C}(x, f) \cap A.$$

Theorem 6.8(a) implies that $\mathcal{C}(x, f) = \mathcal{P}(x, f)$, so if $y \in \omega\mathcal{C}(x, f)$ then there are sequences $x_k \to x$ and $n_k \to \infty$ such that $f^{n_k}(x_k) \to y$. As $X\backslash(A \cup R)$ is open we can assume $x_k \in X\backslash(A \cup R)$ for all k and so $f^{n_k}(x_k) \in X\backslash(A \cup R)$ for all k. Hence, $y \in \partial A$. Similarly, $\omega\mathcal{C}(x, f^{-1}) = \omega(x, f^{-1}) \subset \partial R$.

We have just shown that $x \in X\backslash(A \cup R)$ and $y \in \omega\mathcal{C}(x, f)$ imply $y \in \partial A$. Conversely, if $y \in \partial A$ then there is a sequence $z_k \to y$ with $z_k \in X\backslash A$. As $A \subset U^\circ$ we can assume $z_k \in U$ for all k. So there is a sequence of nonnegative integers n_k such that $x_k = f^{-n_k}(z_k) \in Q$. Furthermore, $n_k \to \infty$ because $\cup_{i=0}^{N} f^i(Q)$ is closed and disjoint from A for any positive integer N. By restricting to a subsequence we can assume $x_k \to x \in Q$. Thus, $y \in A \cap \mathcal{P}(x, f) = A \cap \mathcal{C}(x, f) = \omega\mathcal{C}(x, f)$. This completes the proof that $\partial A = \omega\mathcal{C}(X\backslash(A \cup R), f)$.

To get the rest of (6.19), note that ∂A is closed and invariant, so $\partial A \subset \mathcal{C}(\partial A, f)$. Having a point $y \in \partial A$ with $\mathcal{C}(y, f) \not\subset \partial A$ is impossible by the argument of the last paragraph. This proves the first line of (6.19). The second is the first applied to f^{-1}. In particular, ∂A is a closed $\mathcal{C}(f)$-invariant set and so is a quasi-attractor. Similarly, ∂R is a quasi-repellor.

Now if $x \in R^\circ$ and $y \in \mathcal{C}(x, f) = \mathcal{P}(x, f)$ then there is a sequence $x_k \in R^\circ$ converging to x and a sequence of positive integers n_k such that $f^{n_k}(x_k) \to y$. Since $f(R) = R$, $f^{n_k}(x_k) \in R$ and so $y \in R$. Furthermore, if y were in ∂R then by (6.19) $x \in \mathcal{C}(y, f^{-1})$ would be in ∂R. Hence, $y \in R^\circ$. Thus,

$$R^\circ = f(R^\circ) \subset \mathcal{C}(R^\circ, f) \subset R^\circ, \tag{6.21}$$

and so

$$R^\circ = \mathcal{C}(R^\circ, f). \tag{6.22}$$

If A_1 is an attractor with $A_1 \subset R$ then $W^s_f(A_1)$ is an open set contained in R and so contained in R°. On the other hand, if $x \in R^\circ$ then by (6.23), the closed $\mathcal{C}(f)$-forward invariant set $\mathcal{C}(x, f) \subset R^\circ$. For sufficiently small ϵ, the inward set $U \equiv \mathcal{C}_\epsilon(x, f)$ is contained in R° (see (1.14)). For the associated attractor $A_1 = \omega(U, f)$ we have $A_1 \subset U \subset R^\circ$ and the f-invariant set $\omega\mathcal{C}(x, f)$ is contained in A_1. So x is in the basin $W^s_f(A_1)$. This proves the second half of (6.20). The first follows by applying the second to f^{-1}.

(b): Observe first that $\mathcal{C}(B_1, f) \cap \mathcal{C}(B_2, f^{-1}) = \emptyset$; otherwise, transitivity would imply that $B_2 \subset \mathcal{C}(B_1, f)$, and so the assumption that $B_1 \subset \mathcal{C}(B_2, f)$ would lead to the contradiction that $B_1 \cup B_2$ is contained in a single chain component.

Hence, $\omega\mathcal{C}(B_1, f)$ is a quasi-attractor disjoint from B_2 (see (1.6)). Since the inward sets provide a neighborhood base for a quasi-attractor, we can choose an inward set U with $\mathcal{C}(B_1, f) \subset U$ but $B_2 \cap U = \emptyset$. For $A = \omega(U, f)$ and $R = \alpha(X\backslash U^\circ, f)$, the associated attractor-repellor pair, we have

$$\begin{cases} B_1 \subset A, \quad B_2 \subset R \\ B_1 \cap R = B_2 \cap A = \emptyset. \end{cases} \tag{6.23}$$

If some point z of B_2 were in R° then by (6.20)

$$\mathcal{C}(B_2, f) = \mathcal{C}(z, f) \subset \mathcal{C}(R^\circ, f) \subset R, \tag{6.24}$$

contradicting $B_1 \subset \mathcal{C}(B_2, f)$. Hence, $B_2 \subset \partial R$. Similarly, $B_1 \subset \partial A$.

Conversely, if $B_2 \subset \partial R$ then by (6.19) there exists $x \in X \backslash (A \cup R)$ such that $B_2 \subset \omega\mathcal{C}(x, f^{-1})$. Let $\widetilde{B}_1$ be the chain component containing $\omega(x, f)$. Clearly,

$$\widetilde{B}_1 \subset \omega\mathcal{C}(x, f) \subset \omega\mathcal{C}(B_2, f). \tag{6.25}$$

So $\widetilde{B}_1 \subset \partial A$ by (6.20) again. Applied to f^{-1} this argument constructs $\widetilde{B}_2$ for B_1. □

COROLLARY 6.10. *Assume that for X the collection $\mathcal{U}_F$ of open sets whose closures satisfy the fixed point property, is a basis for X. If $f \in \mathcal{H}_{3,s}[\mathcal{U}_F]$ then the collection of periodic orbits for X is a subset of $CT(f)$ dense with respect to the Hausdorff metric.*

PROOF. If $A \in CT(f)$ and $\epsilon > 0$ there exists an inward set U for f with an $(\epsilon/2, k)$-periodic decomposition of $\mathcal{U}_F$ type, $\{U_i\}$, and such that $d(A, U) < \epsilon/2$. If $x \in U_0$ is a fixed point for f^k then x is a periodic point of period k for f. By Lemma 2.1(a) the orbit of x has Hausdorff distance from U at most $\epsilon/2$. The result follows from the triangle inequality. □

6.5. Shift extensions and the class $\mathcal{H}_4$

We have seen that the adding machines are generic among chain components when the action is strictly spongy. However, other sorts of chain components can occur.

We call a chain component B for f a *shift extension* chain component when it has a k-periodic decomposition $\{B_i\}$ for some k and there is a continuous map π from B_0 onto the two sided symbol space Σ which maps f^k on B_0 to the shift homeomorphism σ on Σ. That is,

$$\pi(f^k(x)) = \sigma(\pi(x)) \quad \text{for } x \in B_0. \tag{6.26}$$

We will obtain shift extension chain components by using topological horseshoes. The results will require special assumptions on H.

For a closed subset K of X and $\epsilon > 0$ we define $G^4(K, \epsilon) \subset H$ as follows:

$$f \in G^4(K, \epsilon) \Leftrightarrow \text{ if either (6.27.1) or (6.27.2) is met:} \tag{6.27}$$

(6.27.1) : $K \cap PNW_H(f) = \emptyset$.

(6.27.2) : There exists an inward set U for f with an (ϵ, k)-periodic decomposition $\{U_i\}$ such that $U_0 \subset V_\epsilon(K)$ and U_0 is a topological horseshoe for f^k.

LEMMA 6.11. *Each $G^4(K, \epsilon)$ is an open subset of H.*

If $\widetilde{K}$ is a closed subset of X such that $K \subset \widetilde{K} \subset V_\epsilon(K)$ then

$$G^4(\widetilde{K}, \epsilon) \subset G^4(K, 2\epsilon). \tag{6.28}$$

Assume that for some space Y, H is a closed subgroup of the automorphism group $H(Y)$ and that the evaluation action of H on Y is strictly spongy. Assume that X is a closed, H invariant subset of Y with only finitely many components. With respect to the action of H on X each $G^4(K, \epsilon)$ is a dense subset of H.

PROOF. If U is inward for f and U_0 is a topological horseshoe for f^k then the same is true for all f_1 close enough to f in H. So openness and (6.28) follow as in Lemma 6.1.

Now given $f \in H$ and $\epsilon_1 > 0$ we will construct $g \in G^4(K, \epsilon)$ ϵ_1-close to f,when the special hypotheses on H and Y are assumed.

By Proposition 4.8(a) the action of H on X is spongy because it was assumed spongy on Y. By Lemma 6.1 we can choose $f_1 \in H$ with $d(f, f_1) < \epsilon_1/2$ and such that $f_1 \in G^2(K, \epsilon, 1)$. Assume that $K \cap PNW_H(f_1) \neq \emptyset$. We first find an f_1-recurrent point in $V_\epsilon(K)$, i.e. find x such that $x \in \omega(x, f_1) \cap V_\epsilon(K)$.

To find x, we use $f_1 \in G^2(K, \epsilon, 1)$ which implies that $V_\epsilon(K)$ contains a piece of a periodic decomposition for some f_1 inward subset. By Lemma 2.1(a) any minimal subset of the associated attractor meets every piece of the decomposition. So we can find a point x in $V_\epsilon(K)$ such that x lies in a minimal subset and so x is recurrent.

Now we apply Proposition 5.4 to the action of H on Y with O_0 the ϵ-neighborhood in Y of K. Since $x \in X$, $\overline{H(x)} \subset X$. By Proposition 4.8(c) applied first to X and then to $\overline{H(x)}$ we see that $\overline{H(x)}$ has only finitely many components.

By Proposition 5.4 there exists $g \in H$ such that $d(f_1, g) < \epsilon_1/2$ and such that the homeomorphism g on Y has an inward set U with an (ϵ, k)-periodic decomposition $\{U_i\}$ such that $U_0 \subset O_0$. Furthermore, there is a succession of subsets $\{V_1, \dots V_5\}$ with respect to which U_0 is a horseshoe for g^k and $\{V_1 \cap \overline{H(x)}, \dots, V_5 \cap \overline{H(x)}\}$ is a connected succession as well. Since $\overline{H(x)} \subset X$, $V_1 \cap X$ and $V_5 \cap X$ are not separated in $U_0 \cap X$ and so $U_0 \cap X$ is a topological horseshoe for the homeomorphism g^k on X with (ϵ, k)-decomposition $\{U_i \cap X\}$. Furthermore, $U_0 \cap X \subset O_0 \cap X = V_\epsilon(K)$. Hence, $g \in G^4(K, \epsilon)$. □

Define yet another subset of H:

$$f \in \mathcal{H}_4 \Leftrightarrow f \text{ satisfies condition 6.29.1:} \tag{6.29}$$

(6.29.1) : For all nonempty closed subsets K of X and $\epsilon > 0$, both f, and f^{-1} are in $G^4(K, \epsilon)$.

THEOREM 6.12. *For the action of the Polish group H on X, the set $\mathcal{H}_4$ is a conjugacy invariant, G_δ subset of H.*

Assume that for some space Y, H is a closed subgroup of the automorphism group $H(Y)$ and that the evaluation action of H on Y is strictly spongy. Assume that X is a closed, H invariant subset of Y with only finitely many components. For the restricted action of H on X, the subset $\mathcal{H}_4$ defined by (6.29) is dense in H.

PROOF. Just as Theorem 6.2 followed from Lemma 6.1, we obtain this result from Lemma 6.11 by an analogous argument. □

THEOREM 6.13. *For the action of H on X, let $f \in \mathcal{H}_4$ as defined by (6.29).*

(a) *The set of shift extension chain components is dense in the space $\mathcal{B}_f$ of all chain components for f.*

(b) *If B is a chain component in $\Pi_\infty(f)$ or $\Pi_\infty(f^{-1})$ then for every $\epsilon > 0$ there is a shift chain component B_1 for f such that the Hausdorff distance $d(B, B_1)$ is less than ϵ.*

PROOF. First assume that $\epsilon > 0$ and $x \in \Omega(f)$. Because $f \in G^4(x, \epsilon)$, there is an inward set U for f with an (ϵ, k)-periodic decomposition $\{U_i\}$ such that

$U_0 \subset V_\epsilon(x)$ and U_0 is a topological horseshoe for f^k. Apply Proposition 5.2 to obtain a chain component B_0 for f^k with $B_0 \subset U_0$, and a surjective continuous map $\pi : B_0 \to \Sigma$ which maps f^k on B_0 to the shift homeomorphism on Σ. Let $B_i = f^i(B_0)$. Since $B_0 \subset U_0$, $B_i \subset U_i$ for all i and so $\{B_i\}$ is a k-periodic decomposition for the f-invariant set B which is its union. Because B_0 is a chain component for f^k, B is a chain component for f by Proposition 1.1. By construction it is a shift chain component.

(a): Begin with any chain component $\widetilde{B}$, any nonwandering point $x \in \widetilde{B}$, and $\epsilon_1 > 0$. Choose $\epsilon > 0$ a modulus of uniform continuity for the complete Lyapunov function L used to define the metric on $\mathcal{B}_f$. From the above construction, we obtain a shift extension chain component B which meets $V_\epsilon(x)$. Hence, $|L(\widetilde{B}) - L(B)| < \epsilon_1$.

(b): If $\widetilde{B} \in \Pi_\infty$ then there exists an $(\epsilon_1/2, k)$-periodic inward set $\widetilde{U}$ with $\widetilde{B} \subset \widetilde{U}$. Supposing x is in the piece $\widetilde{U}_0$ of an (ϵ_1, k) decomposition for $\widetilde{U}$, then $x \in \widetilde{U}_0^\circ$ implies $V_\epsilon(x) \subset \widetilde{U}_0$ for some $\epsilon > 0$. Then the shift extension chain component B constructed above meets $\widetilde{U}_0$ and so is contained in $\widetilde{U}$. By Lemma 2.1(a) $\widetilde{B}$ and B each have Hausdorff distance at most $\epsilon_1/2$ from $\widetilde{U}$.

For $\widetilde{B} \in \Pi_\infty(f^{-1})$ apply the result for f^{-1}. □

COROLLARY 6.14. *If $f \in \mathcal{H}_4 \cap \mathcal{H}_{3,s}$ then the shift extension chain components form a dense subset of $CT(f)$ and their union is a dense subset of $\mathcal{C}(f) = \Omega(f)$.*

PROOF. By Theorem 6.8 (b) the $\Pi_\infty(f)$ chain components are dense in $CT(f)$. So the first result follows from Theorem 6.13 (b). The second result then follows because $\mathcal{C}(f) = \Omega(f)$ by Theorem 6.8 (a) and the $\Pi_\infty(f)$ points are dense in this set by Theorem 6.4 (a). □

We remark in passing, as noted at the very end of chapter 5, it follows that the topologically generic homeomorphism of a manifold has infinite topological entropy.

PROPOSITION 6.15. *Assume that X and Y are compact spaces, and that H and K are Polish spaces, all equipped with metrics (denoted d in each case).*

(a) *With the topology of uniform convergence $C(X; H)$ is a Polish space and the topology is given by the sup metric induced from d on H. The inclusion $H \to C(X; H)$ by constant functions is then isometric.*

(b) *If $g : H \to K$ is continuous then $g_* : C(X, H) \to C(X, K)$ defined by $g_*(F) = g \circ F$ is continuous. Furthermore, the composition and evaluation functions*

$$\text{comp}: \ C(Y; X) \times C(X; H) \to C(Y; H)$$

$$\text{ev}: \ X \times C(X; H) \to H \tag{6.30}$$

are continuous.

(c) *If H is a Polish group then $C(X; H)$ is a Polish group with pointwise multiplication induced from H. The evaluation map ev: $X \times C(X; H) \to H$ is an open, continuous surjection.*

PROOF. The proof is straightforward, and results of this type are common in the literature, so we omit the details. See for example tom Dieck (1987) section I.1; Bourbaki (1966) chapter X §3; Kelley (1955) chapter 7, Michor (1980), §3, Gottschalk and Hedlund (1955), chapter 12. □

THEOREM 6.16. *Assume Y is a compact metric space, H is a Polish group and H^* is a dense, G_δ subset of H. For any $F \in C(Y;H), F^{-1}(H^*)$ is a G_δ subset of Y.*

$$H^*_Y \underset{\text{def}}{=} \{F \in C(Y;H) : F^{-1}(H^*) \text{ is dense in } Y\} \tag{6.31}$$

is a residual subset of $C(Y;H)$, i.e. it contains a dense, G_δ subset of $C(Y;H)$. Furthermore, if H^* is a conjugacy invariant subset of H then H^*_Y is a conjugacy invariant subset of $C(Y;H)$.

PROOF. Because ev is continuous, $\text{ev}^{-1}(H^*)$ is a G_δ subset of $Y \times C(Y;H)$ and similarly each $F^{-1}(H^*)$ is G_δ in Y. Because ev is an open map, H^* dense implies $\text{ev}^{-1}(H^*)$ is dense. Now we apply the *topological Fubini Theorem*, see GTDS Exercise 7.32. This says that if D is completely metrizable, Y is compact metrizable and A is a residual subset of $D \times Y$ then, letting $\pi_S : D \times Y \to S$ ($S = D$ or Y) denote the projections,

$$\{d \in D : \pi_Y(A \cap \pi_D^{-1}(d)) \text{ is a residual subset of } Y\}$$

is residual in D. Applied with $A = \text{ev}^{-1}(H^*)$ it follows that H^*_Y is residual.

For $F \in C(Y;H)$, define $F^- \in C(Y;H)$ by $F^-(y) = (F(y))^{-1}$, i.e., it is the map F followed by inversion in the group. If $F_2 = FF_1F^-$ for $F_1, F, F_2 \in C(Y;H)$ and H^* is conjugacy invariant then for all $y \in Y$, $F_2(y) \in H^*$ iff $F_1(y) \in H^*$, i.e. $F_2^{-1}(H^*) = F_1^{-1}(H^*)$. *A fortiori*, $F_1^{-1}(H^*)$ dense implies $F_2^{-1}(H^*)$ is dense. $\square$

To interpret this result we use the natural isometric identification:

$$C(Y;C(X;H)) \simeq C(Y \times X;H) \tag{6.32}$$

That is, given $F : Y \times X \to H$ continuous then for $y \in Y$, $F_y : X \to H$ is continuous and by uniform continuity F, the map $y \mapsto F_y$ is continuous from Y to $C(X;H)$. Conversely, a continuous map $F : Y \to C(X;H)$ induces a continuous map from $Y \times X \to H$ by successive evaluations.

Now suppose that Polish group H acts on a space X. We can identify $F \in C(Y;H)$ with a homeomorphism $(1,F)$ on $Y \times X$ which projects to 1_Y on Y:

$$(1,F)(y,x) = (y, F(y)(x)). \tag{6.33}$$

On the fiber over y, $(1,F)$ on $\{y\} \times X$ is $F(y)$ on X. When $H = H(X)$, (6.32) says that every homeomorphism of $Y \times X$ which projects to 1_Y can be so obtained from $F \in C(Y;H)$.

If $H^* \subset H$ describes some generic property among the dynamical systems on X described by H, then Theorem 6.16 says that among the parametrized families of $C(Y;H)$ those in the residual subset H^*_Y restrict to H^* systems on a residual set of fibers.

CHAPTER 7

Almost Equicontinuity

Overview: We will show that, under appropriate conditions, a generic homeomorphism has the property that it determines a residual subset of X consisting of points where the map satisfies a condition even stronger than Lyapunov stability (i.e., stronger than equicontinuity at that point).

For a continuous map f on a space X, we call a point $x \in X$ an *equicontinuity point* for f if the sequence of maps $\{f^n : n = 0, 1, \ldots\}$ is equicontinuous at x. Such a point is also called a *Lyapunov stable* point. To be precise, for $\epsilon > 0$ define $Eq(f, \epsilon)$ to be the union of all open subsets O of X with the property that for any $x_1, x_2 \in O$, and $n \geq 0$,

$$d(f^n(x_1), f^n(x_2)) \leq \epsilon.$$

Now define

$$Eq(f) = \bigcap_{\epsilon > 0} Eq(f, \epsilon). \tag{7.1}$$

Each $Eq(f, \epsilon)$ is open and f^{-1}-forward invariant, i.e.

$$f^{-1}(Eq(f, \epsilon)) \subset Eq(f, \epsilon), \tag{7.2}$$

with equality when f is a homeomorphism. Since the sets $Eq(f, \epsilon)$ decrease with ϵ, it follows that the set of equicontinuity points is always a G_δ subset.

Regarded as a dynamical system f is called *equicontinuous* when every point is equicontinuous, i.e. $Eq(f) = X$. In that case, compactness implies *uniform equicontinuity*: for every $\epsilon > 0$, there exists $\delta > 0$ such that $d(x_1, x_2) < \delta$ implies $d(f^n(x_1), f^n(x_2)) \leq \epsilon$ for all integers $n \geq 0$. Following Akin et al. (1996), (1998) we will call f *almost equicontinuous* when $Eq(f)$ is dense in X and so the equicontinuity points are residual in X.

On the other hand, one says that the system exhibits *sensitive dependence on initial conditions*, or simply that f is *sensitive*, when $Eq(f, \epsilon) = \emptyset$ for some $\epsilon > 0$. This says that for some fixed, positive ϵ, every nonempty open subset O contains points x_1, x_2 which are moved at least ϵ apart by some iterate (in fact, infinitely many iterates) of f. Sensitivity is often among the proposed requirements for the label *chaos*.

These concepts have been most studied in the *topologically transitive* case, that is, when f satisfies condition (5.5).

The fundamental result is the *Auslander-Yorke Dichotomy Theorem* (1980):

THEOREM 7.1. *Assume f is a topologically transitive map. If f admits any equicontinuity points then*

$$Eq(f) = \text{Trans}(f) \tag{7.3}$$

and so f is almost equicontinuous. Otherwise f is sensitive.

PROOF. If $Eq(f, \epsilon) \neq \emptyset$ then the orbit of any transitive point x enters this open set. So by (7.2), $x \in Eq(f, \epsilon)$. So if f is not sensitive, i.e. $Eq(f, \epsilon) \neq \emptyset$ for any $\epsilon > 0$, then $\text{Trans}(f) \subset Eq(f, \epsilon)$ for all $\epsilon > 0$. Intersecting over $\epsilon > 0$. We obtain the inclusion $\text{Trans}(f) \subset Eq(f)$ and so f is almost equicontinuous. For the reverse inclusion see Akin et al (1996) Theorem 2.4. □

A topologically transitive, almost equicontinuous map is always a homeomorphism and it is uniformly rigid. A homeomorphism f on X is called *uniformly rigid* if for every $\epsilon > 0$ there is a positive integer n such that

$$d(f^n, 1_X) = d(1_X, f^{-n}) < \epsilon, \tag{7.4}$$

using the uniform metric (3.8) in $H(X)$. (This property has been given several names in the literature; for instance, Furstenberg (1981) calls such a homeomorphism *recurrent*.) We refer to Akin et al (1996) or Glasner and Weiss (1993) for the proof of the following:

THEOREM 7.2. *If f is an almost equicontinuous, topologically transitive map on X, then f is a homeomorphism and f^{-1} is an almost equicontinuous, topologically transitive map, with*

$$Eq(f) = \text{Trans}(f) = \text{Trans}(f^{-1}) = Eq(f^{-1}). \tag{7.5}$$

Furthermore, f is uniformly rigid. □

For a homeomorphism f on $H(X)$ we can define the smallest closed subgroup of $H(X)$ which contains f by

$$\Gamma_f = \overline{\{f^n : n \in \mathbf{Z}\}}. \tag{7.6}$$

A pair (Γ, g) where $g \in \Gamma$ is called a *monothetic group with generator* when Γ is a topological group in which the cyclic group generated by g is dense. Because Γ is the closure of an abelian subgroup, it is abelian.

So (Γ_f, f) is a monothetic group with generator. When the sequence $\{f^n : n \geq 0\}$ in $H(X)$ is discrete then Γ_f is just the discrete cyclic group $\{f^n : n \in \mathbf{Z}\}$. f is uniformly rigid exactly in the complementary case when the sequence $\{f^n : n \geq 0\}$ has a nonempty set of limit points in $H(X)$.

If a mapping f on X is minimal then $\text{Trans}(f) = X$. So Theorem 7.1 implies that an almost equicontinuous minimal system is equicontinuous. It is a classic theorem of Gottschalk and Hedlund (1955) that such a system is always the translation by a generator of a compact monothetic group.

THEOREM 7.3. *Assume that f is a minimal homeomorphism on X and that f is not sensitive. Let $x \in X$. The system f is equicontinuous and Γ_f is a compact monothetic group with generator f. The evaluation map $e_x : \Gamma_f \to X$ defined by $e_x(g) = g(x)$ is a homeomorphism providing a conjugacy from translation by f on Γ_f to the homeomorphism f itself on X.*

PROOF. The set $\{f^n : n \in \mathbf{Z}\}$ is uniformly equicontinuous because f^{-1} is equicontinuous as well as f. Hence, Γ_f is compact by the Arzela-Ascoli Theorem. Since the orbit of x is dense in X, e_x is surjective. If $g_1(x) = g_2(x)$ for $g_1, g_2 \in \Gamma_f$, then, because Γ_f is abelian,

$$g_1(f^n(x)) = f^n(g_1(x)) = f^n(g_2(x)) = g_2(f^n(x)), \tag{7.7}$$

for all n and so $g_1 = g_2$ by continuity. By compactness, e_x is a homeomorphism. The remaining result is clear. □

The converse result, that translation on a compact abelian group is an equicontinuous homeomorphism, is obvious because such a translation is an isometry of any invariant metric.

A minimal map on a finite set is just a single periodic orbit which is conjugate to translation by a generator in a finite cyclic group. On the other hand, a homeomorphism f on X was defined above to be an *adding machine* when it is conjugate to translation by a generator in an infinite group which is the inverse limit of finite cyclic groups; see the discussion preceding Proposition 2.2.

PROPOSITION 7.4. *Let f be a homeomorphism on X. The system (X, f) is an adding machine iff f is a minimal, equicontinuous homeomorphism and X is an infinite, zero-dimensional space.*

PROOF. This reduces to showing that a zero-dimensional compact monothetic group is an inverse limit of finite cyclic groups. See, e.g. Akin (1996) Corollary 3.2 and Theorem 3.5. □

7.1. Chain continuity

It is easy to see that x is an equicontinuity point for a map f iff for every $\epsilon > 0$ there exists $\delta > 0$ such that for all integers $n \geq 0$ and points $x_1 \in X$, $d(x, x_1) \leq \delta$ implies $d(f^n(x), f^n(x_1)) \leq \epsilon$. We define x to be a *chain continuity point* for f if for every $\epsilon > 0$ there exist a $\delta > 0$ such that if $\{x_0, \dots, x_n\}$ is a sequence in X:

$$\begin{cases} d(x, x_0) \leq \delta \text{ and } d(f(x_{i-1}), x_i) \leq \delta \; i = 1, \dots, n \\ \\ \qquad \Rightarrow d(f^i(x), x_i) \leq \epsilon \quad i = 0, \dots, n. \end{cases} \tag{7.8}$$

In words: every δ-chain which starts δ-close to x *ϵ-shadows* the orbit of x.

Chain continuity was introduced in Akin (1996). It is such a strong condition that it is possible to completely characterize when it occurs.

THEOREM 7.5. *Let f be a continuous map on a space X and $x \in X$. The following are equivalent:*

(1) *x is a chain continuity point for f.*

(2) *$\omega\mathcal{C}(x, f) = \omega(x, f)$ and the set is either a periodic orbit or an adding machine subset for f.*

(3) *$x \in W^s_f(\Pi_\infty(f))$.*

PROOF. By Akin (1996) Theorem 3.3 and Corollary 3.2, x is a chain continuity point for f iff $\omega(x, f) = \omega\mathcal{C}(x, f)$ and this subset is a zero-dimensional invariant set on which f restricts to a minimal equicontinuous homeomorphism. By Proposition 7.4 this says exactly that $\omega(x, f)$ is either a periodic orbit or an adding machine subset for f. Thus, (1) $\Leftrightarrow$ (2). By Lemma 1.5 (d), $\omega(x, f) = \omega\mathcal{C}(x, f)$ iff $\omega(x, f)$ is a terminal chain component. If (2) holds then by Proposition 2.2 the points of $\omega(x, f)$ lie in $\Pi_\infty(f)$ and so (3) is true. The converse, (3) $\Rightarrow$ (2), follows from Corollary 2.4. □

THEOREM 7.6. *For the action of a Polish group H on the space X, assume $f \in \mathcal{H}_1$ as defined by (6.5). The set $W^s_f(\Pi_\infty(f))$ of chain continuity points for f,* *the set* $W^u_f(\Pi_\infty(f^{-1}))$ *of chain continuity points for* f^{-1}, *and their intersection are* G_δ *sets dense in* X.

$$W^s_f(\Pi_\infty(f)) \cap W^u_f(\Pi_\infty(f^{-1})) \cap \mathcal{C}f = \Pi_\infty(f) \cap \Pi_\infty(f^{-1}) \tag{7.9}$$

is the set of chain recurrent points on which f *and* f^{-1} *are both chain continuous. This is a* G_δ *subset dense in* $\mathcal{C}(f)$.

PROOF. The identification of $W^s_f(\Pi_\infty(f))$ as the set of chain continuity points for f is exactly the above theorem. It is a dense G_δ subset of X by Theorem 6.4 (b). If $x \in \mathcal{C}(f) \cap W^s(\Pi_\infty(f))$ then the chain component B that contains $\omega(x, f)$ is either an attractor or a quasi-attractor by Proposition 2.3. Consequently, 1.11 shows that x being chain recurrent implies that $x \in B$, and so $x \in \Pi_\infty(f)$. That is

$$W^s_f(\Pi_\infty(f)) \cap \mathcal{C}(f) = \Pi_\infty(f). \tag{7.10}$$

Together with the analogous result for f^{-1} this proves equation (7.9). Density in $\mathcal{C}(f)$ then follows from Theorem 6.4 (a). □

COROLLARY 7.7. *The homeomorphisms of* $\mathcal{H}_1$ *are almost equicontinuous on the space* X.

In these cases it is easy to see how chain continuity occurs. Suppose that U is an inward set for f with an (ϵ, k)-periodic decomposition $\{U_i\}$. Because U is inward, there exists $\delta_1 > 0$ so that

$$\overline{V_{\delta_1}(f(U_i))} \subset U_{i+1} \tag{7.11}$$

for all i. If $\overline{V_{\delta_1}(x)} \subset U_0$, e.g. if $x \in f^k(U_0)$, then for any δ_1-chain $\{x_0, \ldots, x_n\}$ which starts δ_1 close to x, we have $x_i \in U_i$ for all $i \geq 0$. So $d(f^i(x), x_i) \leq \epsilon$ for all $i \geq 0$ because the diameter of U_i is at most ϵ. If $x \in W^s(U)$ then for some n_0 we have $f^{n_0}(x) \in f^k(U_0)$. By uniform continuity of $f, \ldots, f^{n_0}$ we can choose δ_2 so that if $\{x_0, \ldots, x_n\}$ is a δ_2 chain starting δ_2 close to x, then $d(f^i(x), x_i) \leq \delta_1$ for all i such that $0 \leq i \leq n_0$. The previous result applies for $n_0 \leq i \leq n$. So we then get $d(f^i(x), x_i) \leq \epsilon$ for all i up to n.

What the above result proves is that for any x in the open set

$$W^s(\bigcup_k \Pi_{(n,k)}(f))$$

there exists $\delta(x) > 0$ so that any $\delta(x)$-chain which begins $\delta(x)$-close to x is $(1/n)$-shadowed by the positive orbit of x. When $f \in \mathcal{H}_1$ this open subset is dense in X. The chain continuity points are obtained by intersecting over n.

In contrast with the topologically transitive case, points which are Lyapunov stable or chain continuous for f need not satisfy the same condition for f^{-1}.

Suppose that $B_1 \subset \Pi_\infty(f)$ and $B_2 \subset \Pi_\infty(f^{-1})$ are distinct chain components. The points of $W^s_f(B_1)$ are chain continuity points for f and $W^u_f(B_2)$ are chain continuity points for f^{-1}. If $W^s_f(B_1) \cap W^u_f(B_2) \neq \emptyset$ then this set of chain continuity points for f and f^{-1} excludes both B_1 and B_2. If x is in the intersection then $\omega(x, f) = B_1$ and $\alpha(x, f) = B_2$. We see that the points of B_2 are not Lyapunov stable for f because there are points on the orbit of x which are arbitrarily close to any point in B_2. While B_2 is a closed, invariant set, the points on the orbit of x move away towards B_1. Similarly, the points of B_1 are not equicontinuity points for f^{-1}.

Here is an analogy that may help explain how the generic homeomorphism can have chaotic behavior – the topological horseshoes – and yet be chain continuous at a residual subset of X.

Recall from Corollary 6.14 and Theorem 6.7 that if $f \in \mathcal{H}_{3,s} \cap \mathcal{H}_4 \subset \mathcal{H}_1$ then f contains infinitely many horseshoes leading to shift extension chain components. This implies that such homeomorphisms have positive topological entropy. The explanation of the coexistence of positive entropy with almost equicontinuity is quite simple – the topological horseshoes are of first Baire category, missing the residual set of points where we have chain equicontinuity. In the next paragraph we give another example of this type of phenomenon, but in a simpler setting.

Suppose you begin with the original example of a Morse function on the torus from Milnor (1963) and let f be the time one homeomorphism for the associated gradient vector field. The chain recurrent set consists of four fixed points: a repellor, an attractor and two saddles. The closure of the stable and unstable manifolds of the saddles consists of four closed arcs. On the remaining open dense set every point x is chain continuous for f and f^{-1} with $f^n(x)$ moving toward the attracting sink as $n \to \infty$ and back to the repelling source as $n \to -\infty$. Once you leave behind equicontinuous minimal systems, the dynamics are about as regular as you can get.

Now perturb f in the neighborhood of one of the saddle points by introducing a small S-shaped Smale-type hyperbolic horseshoe. The saddle point of f is replaced by a small Cantor set on which the new homeomorphism f_1 restricts to a full shift on 3 symbols. So f_1 has positive entropy. Nonetheless, it is still true that except for points on a closed, one dimensional, nowhere dense subset every orbit proceeds from repellor to the attractor in a chain continuous fashion. Near the Cantor set the orbits jump around a bit but then they leave and move uneventfully to the attractor. Our $\mathcal{H}_1$ homeomorphisms behave in a similar way, even though they exhibit other much more complicated behavior. Being able to make our perturbations in the topological category means that we have relatively few obstructions, and using this freedom to put lots of adding machine sets into the dynamical picture translates into the fact that the equicontinuous dynamics are the topologically typical ones.

CHAPTER 8

Cantor Sets

Overview: Here we illustrate what our results imply about a space quite unlike a manifold, namely a Cantor set – a space that is perfect, and zero-dimensional in addition to being a compact metric space. Dynamics on Cantor sets have been heavily studied, and so many of the techniques we employ here are well known; we believe, however, that our main results, Theorems 8.3, 8.4, 8.10, and 8.11 are new. Other descriptions of generic properties of homeomorphisms of Cantor sets can be found in Sears (1971, 1972).

We describe the generic homeomorphisms on a Cantor set, showing in particular that they have no periodic points. we also show that they have a certain 'universal' property: given any positive integer n and any subset P of $\{1, 2, \dots, n\} \times \{1, 2, \dots, n\}$ such that for each i there are j and k with (i, j) and (k, i) in P, there is a partition of the Cantor set into n closed sets that can be labelled X_i, $1 \leq i \leq n$, with the property that $(i, j) \in P$ if and only if $f(x_i) \cap X_j \neq \emptyset$. We also show the existence of a residual subset of all homeomorphisms with the property that f is in this subset if and only if the topological conjugacy class of f is dense in the set of all homeomorphisms.

A Cantor set has a rich supply of homeomorphisms because of a fact which we will refer to as the *Uniqueness of Cantor*: any two Cantor sets are homeomorphic (see Hocking and Young, 1955, p. 100). In particular, any two nonempty clopen subsets of a Cantor set are homeomorphic. Using this fact we have already seen that a Cantor set is strictly spongy, see Proposition 4.6 (d). Similarly, a Cantor set is easily proved to be generalized homogeneous, see e.g. GTDS Exercise 9.16. We speak of the space being spongy or generalized homogeneous because we are restricting attention in this section to the evaluation action, i.e. to the $H = H(X)$ case.

8.1. Aperiodicity and the class $\mathcal{H}_5$

We first show that, in contrast with the connected case in Corollary 6.5, here it is "no periodic points" which is the generic case. While it seems to us unlikely that this fact had not been known long before this, we do not know of a reference. For any map g on X, let $\text{Fix}(g)$ be the set of fixed points of g. Define for any integer n

$$G^5(n) = \{f : \text{Fix}(f^n) = \emptyset\}. \tag{8.1}$$

PROPOSITION 8.1. *For any positive integer n, $G^5(n)$ is an open, conjugacy invariant subset of $H(X)$. If X is a Cantor set then each $G^5(n)$ is dense in $H(X)$.*

PROOF. By compactness, $\text{Fix}(f^n) = \emptyset$ if and only if the nonnegative function $x \mapsto d(f^n(x), x)$ is bounded away from 0. This is clearly an open condition on f.
Since $h(\text{Fix}(f^n)) = \text{Fix}((hfh^{-1})^n)$, the set $G^5(n)$ is clearly conjugacy invariant.

While one can prove directly that the sets $G^5(n)$ are dense in the Cantor set case, the argument is somewhat lengthy. Instead, we will complete the proof later by obtaining the result as a corollary of Theorem 8.4 below. □

Define

$$\mathcal{H}_5 = \bigcap_n G^5(n); \tag{8.2}$$

we can summarize our discussion as

$$\mathcal{H}_5 \text{ is a dense } G_\delta \text{ and } f \in \mathcal{H}_5 \Rightarrow \mathrm{Per}(f) = \emptyset. \tag{8.3}$$

8.2. The class $\mathcal{H}_6$

Because of the rich supply of partitions on a Cantor set, there are special generic properties for homeomorphisms of Cantor spaces.

Recall that a *partition* $\mathcal{A}$ on a space X is a finite, pairwise disjoint cover of X by nonempty clopen subsets. For a positive integer n, let

$$[n] \underset{\text{def}}{=} \{1, \ldots, n\}.$$

We will refer to a continuous surjective map from X to some $[n]$ as a *numbered partition*. If $\alpha : X \to [n]$ is a numbered partition then

$$\mathcal{A}^\alpha \underset{\text{def}}{=} \{\alpha^{-1}(i) : i \in [n]\} \tag{8.4}$$

is the associated partition proper. Each $\alpha^{-1}(i)$ is nonempty because α is surjective and clopen because α is continuous with $[n]$ discrete.

A partition $\mathcal{A}$ determines an equivalence relation, which we denote $\sim_\mathcal{A}$, whose equivalence classes are just the elements of $\mathcal{A}$. There is an induced clopen equivalence relation on $H(X)$, also denoted $\sim_\mathcal{A}$.

$$f_1 \sim_\mathcal{A} f_2 \Leftrightarrow f_1(x) \sim_\mathcal{A} f_2(x) \text{ for all } x \in X. \tag{8.5}$$

Using the change of variable $x \mapsto h(x)$ we see that for any $h \in H(X)$

$$f_1 \sim_\mathcal{A} f_2 \Leftrightarrow f_1 \circ h \sim_\mathcal{A} f_2 \circ h. \tag{8.6}$$

For a numbered partition, α, we write the associated equivalence relation $\sim_\alpha$. So $x_1 \sim_\alpha x_2$ iff $\alpha(x_1) = \alpha(x_2)$. Hence

$$f_1 \sim_\alpha f_2 \Leftrightarrow \alpha \circ f_1 = \alpha \circ f_2; \tag{8.7}$$

in what follows we will abbreviate $\alpha \circ f$ to αf.

Given a surjective continuous map f on X and a partition $\mathcal{A}$ on X, the *pullback* is

$$f^{-1}\mathcal{A} \underset{\text{def}}{=} \{f^{-1}(A) : A \in \mathcal{A}\}. \tag{8.8}$$

If $\alpha : X \to [n]$ is a numbered partition then

$$f^{-1}\mathcal{A}^\alpha = \mathcal{A}^{\alpha f} \tag{8.9}$$

or, equivalently, for $\alpha, \beta : X \to [n]$,

$$\beta = \alpha f \Leftrightarrow f^{-1}(\alpha^{-1}(i)) = \beta^{-1}(i) \text{ for all } i \in [n]. \tag{8.10}$$

Given two partitions $\mathcal{A}_1, \mathcal{A}_2$ then their *join* is

$$\mathcal{A}_1 \wedge \mathcal{A}_2 \underset{\text{def}}{=} \{A_1 \cap A_2 : A_1 \in \mathcal{A}_1 \text{ and } A_2 \in \mathcal{A}_2\} \backslash \{\emptyset\}. \tag{8.11}$$

Given two numbered partitions $\alpha : X \to [n]$ and $\beta : X \to [m]$ we define the join

$$\begin{cases} \alpha \otimes \beta : X \to [n] \times [m] \text{ by} \\ \alpha \otimes \beta(x) = (\alpha(x), \beta(x)). \end{cases} \tag{8.12}$$

Since $\alpha \otimes \beta$ is usually not surjective and the range is not of the form $[k]$ for a positive integer k, this is not a numbered partition. However, we do have for $(i, j) \in [n] \times [m]$

$$(\alpha \otimes \beta)^{-1}(i, j) = \alpha^{-1}(i) \cap \beta^{-1}(j) \tag{8.13}$$

and so, as collections of subsets of X, we have

$$\mathcal{A}^\alpha \wedge \mathcal{A}^\beta = \{(\alpha \otimes \beta)^{-1}(i, j) : (i, j) \in [n] \times [m]\} \backslash \{\emptyset\}. \tag{8.14}$$

If $\alpha : X \to [n]$ is a numbered partition and if f is a continuous map on X then f^α is defined to be the set

$$f^\alpha = \{\ (\alpha(x), \alpha(f(x)))\ :\ x \in X\ \} \subset [n] \times [n]. \tag{8.15}$$

Note that f^α is a relation on $[n]$ (i.e., a subset of $[n] \times [n]$). Given a homeomorphism $f \in H(X)$ and a relation φ on $[n]$ we say that f *represents* φ *via* α if $f^\alpha = \varphi$. Since f is a homeomorphism this requires that φ be a *surjective relation* on $[n]$, meaning that $\pi_1^{-1}(x) \cap \varphi \neq \emptyset$ and $\pi_2(\varphi) = [n]$, where the π_i's denote the two projections $[n] \times [n] \to [n]$. We will show that a generic element of $H(X)$ represents every surjective relation.

It will be useful to observe that the relation f^α is the image of the map $\alpha \otimes (\alpha f)$, as defined by (8.12).

LEMMA 8.2. *If $\alpha, \beta : X \to [n]$ are numbered partitions and $f, g \in H(X)$ for a Cantor space X, then $f^\beta = g^\alpha$ iff there exists $h \in H(X)$ such that*

$$\beta = \alpha h \text{ and } g \sim_\alpha hfh^{-1}. \tag{8.16}$$

If such an h exists, then $\beta \otimes (\beta f) = (\alpha \otimes (\alpha g)) \circ h$.

PROOF. First assume that that there is an h as in (8.16). By (8.7), $\alpha g h = \alpha h f = \beta f$. Hence, $\beta \otimes \beta f = \alpha h \otimes \alpha g h = (\alpha \otimes \alpha g) \circ h$. It follows that the maps $\beta \otimes \beta f$ and $\alpha \otimes \alpha g$ have the same images, which is to say that $f^\beta = g^\alpha$.

For the converse, assume $f^\beta = g^\alpha$. This means that $(i, j) \in f^\beta$ if and only if $(i, j) \in g^\alpha$. In other words, the two partitions $\{\alpha^{-1}(i) \cap g^{-1}(\alpha^{-1}(j))\}$ and $\{\beta^{-1}(i) \cap f^{-1}(\beta^{-1}(j))\}$ are indexed by exactly the same pairs (i, j) in φ. The pieces of these partitions are clopen, so by uniqueness of Cantor we can relate the corresponding clopen sets by homeomorphisms thus building $h \in H(X)$ such that

$$(\beta(x), \beta(f(x))) = (i, j) \Leftrightarrow (\alpha(h(x)), \alpha(g(h(x)))) = (i, j). \tag{8.17}$$

This means that we have equality of maps $\beta = \alpha h$ and $\beta f = \alpha g h$. Using the first of these equalities in the second, and pre-composing with h^{-1}, we see that $\alpha h f h^{-1} = \alpha g$, so $hfh^{-1} \sim_\alpha g$, and we have (8.16). □

In the following definition, $\varphi \subset [n] \times [n]$ is a particular surjective relation (as defined above), and α denotes a numbered partition, $\alpha : X \to [n]$. Define

$$G^6(\varphi) \underset{\text{def}}{=} \{f \in H(X) : \text{ there is } \alpha \text{ with } f^\alpha = \varphi\}. \tag{8.18}$$

THEOREM 8.3. *For a Cantor set X, each $G^6(\varphi)$ is a dense, open, conjugacy invariant subset of $H(X)$.*

PROOF. If $f_1 \sim_\alpha f$ then $\alpha f_1 = \alpha f$ and so $f^\alpha = f_1^\alpha$, which shows that if f represents φ, then so do all f_1's in the clopen $\sim_\alpha$ equivalence class of f, using the same numbered partition α. Hence, $G^6(\varphi)$ is open.

By Lemma 8.2, if $g = hfh^{-1}$ and $f^\beta = \varphi$ then setting $\alpha = \beta h^{-1}$ gives $g^\alpha = \varphi$. Hence, $G^6(\varphi)$ is conjugacy invariant.

Now given arbitrary $f \in H(X)$ and $\beta : X \to [m]$ we will construct $\alpha : X \to [n]$ and $g \in H(X)$ such that:

$$g \sim_\beta f \quad \text{and} \quad g^\alpha = \varphi. \tag{8.19}$$

If β is chosen so that the diameter $\beta^{-1}(k)$ is less than ϵ for all $k \in [m]$ then $g \sim_\beta f$ implies $d(g, f) < \epsilon$.

Let $\mathcal{B} \equiv \mathcal{A}^\beta$ denote the partition induced by β (see (8.4)), and let $\mathcal{B}'$, $\mathcal{B}''$ denote the partitions $\mathcal{B} \wedge f^{-1}\mathcal{B}$ and $f\mathcal{B} \wedge \mathcal{B} \wedge f^{-1}\mathcal{B}$, respectively. For each $K \in \mathcal{B}''$, choose $\alpha : K \to [n]$ surjective. Assembling these α's together we obtain a numbered partition $\alpha : X \to [n]$ which we think of as *transverse* to $\mathcal{B}''$.

For $1 \leq m \leq n$, define

$$V_m = \{(i, j) \in \varphi \ : \ i = m \},$$
$$H_m = \{(i, j) \in \varphi \ : \ j = m \}.$$

The hypotheses on φ ensure that each of these sets is nonempty. For each $i, j \in [n]$ and element K of $\mathcal{B}''$ choose continuous surjections:

$$\begin{cases} \delta_i : K \cap \alpha^{-1}(i) \longrightarrow V_i \\ \\ \rho_j : K \cap \alpha^{-1}(j) \longrightarrow H_j. \end{cases} \tag{8.20}$$

Putting them together we define continuous surjections

$$\begin{cases} \qquad\qquad \delta, \rho : K \longrightarrow \varphi \\ \\ \delta(x) = (\alpha(x), \delta_{\alpha(x)}(x)) \quad \text{and} \quad \rho(x) = (\rho_{\alpha(x)}(x), \alpha(x)). \end{cases} \tag{8.21}$$

Now we will define g on the pieces of the partition $\mathcal{B}' = \mathcal{B} \wedge f^{-1}\mathcal{B}$.

If $B_1 \cap f^{-1}(B_2) \in \mathcal{B}'$ then because this set is partitioned by members of $\mathcal{B}''$, δ is surjective on it. Its image under f is $f(B_1) \cap B_2$ which is also partitioned members of $\mathcal{B}''$ and so ρ is surjective on this set. We define g mapping the δ partition on $B_1 \cap f^{-1}(B_2)$ to the ρ partition of $f(B_1) \cap B_2$. That is, for each $(i, j) \in \varphi$ define the homeomorphism g by uniqueness of Cantor so that

$$g(\delta^{-1}(i, j) \cap B_1 \cap f^{-1}(B_2)) = \rho^{-1}(i, j) \cap f(B_1) \cap B_2. \tag{8.22}$$

In particular, f and g both map the points of $B_1 \cap f^{-1}(B_2)$ into $f(B_1) \cap B_2$. Hence, $f(x) \in B_2$ implies $g(x) \in B_2$ for all $x \in X$. Thus, $g \sim_{\mathcal{B}} f$ or equivalently $g \sim_\beta f$.

For any $x \in X$, there is a unique $B_1, B_2 \in \mathcal{B}$ and $(i,j) \in \varphi$ such that $x \in \delta^{-1}(i,j) \cap B_1 \cap f^{-1}(B_2)$. By (8.22) $g(x) \in \rho^{-1}(i,j)$. By (8.21) $\alpha(x) = i$ and $\alpha(g(x)) = j$, i.e. $(\alpha \otimes \alpha g)(x) = (i,j)$. Since every $(i,j) \in \varphi$ is thus realized by some x in $B_1 \cap f^{-1}(B_2)$, we have $g^\alpha = \varphi$, proving (8.19). □

Now we define

$$\mathcal{H}_6 \underset{\text{def}}{=} \bigcap_{\varphi} G^6(\varphi), \tag{8.23}$$

where the intersection is taken over all surjective relations φ on $[n]$, for all n. For a given n, such a relation is a subset of $[n] \times [n]$, so $\mathcal{H}_6$ is residual by Theorem 8.3.

Observe that for any numbered partition $\alpha : X \to [n]$

$$(i,j) \in (f^{-1})^\alpha \Leftrightarrow (j,i) \in f^\alpha. \tag{8.24}$$

Consequently f^{-1} represents every surjective relation if f does:

$$f \in \mathcal{H}_6 \Leftrightarrow f^{-1} \in \mathcal{H}_6. \tag{8.25}$$

8.3. Rohlin property

We use $\mathcal{H}_6$ to prove that a Cantor set X has the *Rohlin Property.* This says that X admits a homeomorphism f whose conjugacy class is dense in the entire automorphism group. A weaker result can be found in Sears (1971), where it is shown that there is a homeomorphism whose conjugacy class is dense in the set of all homeomorphisms with the topology of pointwise convergence.

THEOREM 8.4. *If X is a Cantor set then $\mathcal{H}_6$ defined by* (8.23) *is a dense, G_δ subset of $H(X)$ which is conjugacy invariant. Furthermore, f lies in $\mathcal{H}_6$ iff its conjugacy class is dense in $H(X)$. That is,*

$$f \in \mathcal{H}_6 \Leftrightarrow H(X) = \overline{\{hfh^{-1} : h \in H(X)\}}. \tag{8.26}$$

PROOF. By Theorem 8.3, each $G^6(\varphi)$ is open, dense and conjugacy invariant. Hence, $\mathcal{H}_6$ is residual as remarked following (8.23).

Now suppose that the conjugacy class of f is dense. For any surjective relation φ on $[n]$, $G^6(\varphi)$ is open and nonempty. So the conjugacy class of f meets $G^6(\varphi)$. Since $G^6(\varphi)$ is conjugacy invariant $f \in G^6(\varphi)$. Since φ was arbitrary $f \in \mathcal{H}_6$.

On the other hand, assume $f \in \mathcal{H}_6$ and $g \in H(X)$. Given $\epsilon > 0$ let $\alpha : X \to [n]$ be any numbered partition of *mesh* less than ϵ, i.e. diam $\alpha^{-1}(i) < \epsilon$ for $i \in [n]$. Let $\varphi = g^\alpha$. Since $f \in G^6(\varphi)$, there exists $\beta : X \to [n]$ such that $f^\beta = \varphi = g^\alpha$. By Lemma 8.2, there exists $h \in H(X)$ such that with $\beta = \alpha h$, $g \sim_\alpha hfh^{-1}$. Hence, $d(g, hfh^{-1}) < \epsilon$. □

COROLLARY 8.5. *If X is a Cantor set then $\mathcal{H}_6$ is a subset of the set $\mathcal{H}_5$ defined in* (8.2)*. It follows that $\mathcal{H}_5$ is a dense, G_δ subset of $H(X)$ that is conjugacy invariant (this completes the proof of Proposition 8.1). In particular, if $f \in \mathcal{H}_6$ then f has no periodic points.*

PROOF. By Proposition 8.1, $G^5(n)$ is open, and it is trivial to see that it is nonempty (for some prime $p > n$ decompose X into p disjoint clopen sets; by the uniqueness of Cantor all of these pieces are homeomorphic, so there is a homeomorphism of X that permutes these pieces cyclically, so that any periodic point has period a multiple of p). So the dense conjugacy class of f meets $G^5(n)$. Since $G^5(n)$ is conjugacy invariant, $f \in G^5(n)$. Hence, for all n, $\mathrm{Fix}(f^n) = \emptyset$. □

REMARK 8.1. This argument shows that if G is any open, nonempty conjugacy invariant subset of $H(X)$ then $\mathcal{H}_6 \subset G$. This in turn is a specialization to the adjoint action of $H(X)$ on itself that – for any topologically transitive group action on a Polish space – the set of transitive points is contained in any nonempty, invariant open subset.

REMARK 8.2. Sears (1972) contains a proof that the set of expansive homeomorphisms of the Cantor set is dense, but of the first category. (The definition of *expansive homeomorphism* can be found below, (11.5).) Notice that an expansive homeomorphism has no adding machine points, so on the Cantor set, any $f \in \mathcal{H}_1 \cap \mathcal{H}_5$ is not expansive.

Actually, $\mathcal{H}_6$ satisfies a special condition much stronger than conjugacy invariance.

PROPOSITION 8.6. *Assume X is a Cantor set, $f_1, f_2 \in H(X)$ and $\pi : X \to X$ is a continuous surjection mapping f_1 to f_2, i.e.*

$$\pi \circ f_1 = f_2 \circ \pi. \tag{8.27}$$

If $f_2 \in \mathcal{H}_6$ then $f_1 \in \mathcal{H}_6$.

PROOF. Assume $\alpha = X \to [n]$ is a numbered partition.

$$(\alpha \otimes \alpha f_2)\pi = (\alpha\pi) \otimes (\alpha f_2 \pi) = (\alpha\pi) \otimes (\alpha\pi f_1). \tag{8.28}$$

Since $\pi(X) = X$, (8.15) implies that with $\beta = \alpha\pi$,

$$f_1^\beta = f_2^\alpha. \tag{8.29}$$

Hence, if $f_2 \in G^6(\varphi)$ then $f_1 \in G^6(\varphi)$. □

Notice that if X is a Cantor set then $X \times X$ is a Cantor set and so by uniqueness of Cantor there exists a homeomorphism X onto $X \times X$. We will call such a homeomorphism a *squaring homeomorphism*. We use squaring homeomorphisms to build a rich supply of elements of $\mathcal{H}_6$.

COROLLARY 8.7. *Assume X is a Cantor set and $s : X \to X \times X$ is a squaring homeomorphism. If $f \in \mathcal{H}_6$ and $h \in H(X)$ then $s^{-1}(f \times h)s \in \mathcal{H}_6$.*

PROOF. Define $\pi : X \to X$ to be the composition $\pi_1 s$ where $\pi_1 : X \times X \to X$ is the projection on the first coordinate. Clearly, s maps $s^{-1}(f \times h)s$ to $f \times h$ on $X \times X$ and π_1 maps $f \times h$ to f. The result follows from Proposition 8.6. □

8.4. The class $\mathcal{H}_{3,c}$

For a Cantor set we can sharpen some of the constructions from earlier sections.

LEMMA 8.8. *For $f \in H(X)$ and $U \subset X$ the following are equivalent:*

(1) *U is clopen and f-invariant.*

(2) *U is inward for f and for f^{-1}.*

(3) *U is an attractor for f with $W_f^s(U) \backslash U = \emptyset$.*

PROOF. Observe that U is clopen iff $U \subset\subset U$. So (1), i.e. U is clopen and $f(U) = U$ (and hence $U = f^{-1}(U)$), implies U is inward for f and for f^{-1}, with associated attractor $\omega(U, f) = U$, and dual repellor $X \backslash U$. Hence, (1) $\Rightarrow$ (2) and (3). On the other hand, $f(U) \subset\subset U$ and $f^{-1}(U) \subset\subset U$ imply $f(U) = U \subset\subset U$, and so (2) $\Rightarrow$ (1). Finally, if U is an attractor with empty proper basin then $X \backslash U$ is the dual repellor and $\{U, X \backslash U\}$ is a partition of X. Hence, (3) $\Rightarrow$ (1). □

For $f \in H(X)$ and positive integers n, let $\mathcal{W}(f, n)$ denote the collection of all $(1/n, k)$-periodic clopen f-invariant subsets of X (for any $k \geq 1$), and define the f-invariant set:

$$\Pi_{\infty,c}(f) = \bigcap_{n} \bigcup_{U \in \mathcal{W}(f,n)} U \tag{8.30}$$

From Lemma 8.8 and the definitions (2.7) and (8.30):

$$\Pi_{\infty,c}(f) \subset \Pi_{\infty}(f) \cap \Pi_{\infty}(f^{-1}). \tag{8.31}$$

Hence, from Proposition 2.2 we have

$$\Pi_{\infty,c}(f) \cap \mathrm{Fix}(f) = \Pi_{\infty,c}(f) \backslash \Pi_{\infty,\infty}(f). \tag{8.32}$$

Just as in Proposition 2.2, if $x \in \Pi_{\infty,c}(f)$ then $\omega(x, f) \subset \Pi_{\infty,c}(f)$ and so $\omega(x, f)$ is a chain component which is both terminal and initial. So if B is a chain component which meets, and so is contained in $\Pi_{\infty,c}(f)$, then B is dynamically isolated and as in (6.18), $\mathcal{C}(B, f) = B = \mathcal{C}(B, f^{-1})$.

For K a closed subset of X, and $\epsilon > 0$, define a subset $G_c^3(K, \epsilon) \subset H$ by:

$$f \in G_c^3(K, \epsilon) \Leftrightarrow f \text{ satisfies property (8.33.1) :} \tag{8.33}$$

(8.33.1) : There exists U an (ϵ, k) periodic clopen, f-invariant subset of X, for some integer k, with $d(K, U) < \epsilon + d(K, A)$ for all $A \in CT(f)$.

LEMMA 8.9. *For a Cantor set X, each $G_c^3(K, \epsilon)$ is a dense open subset of $H(X)$. If $\widetilde{K}$ is a closed subset of X such that $K \subset \widetilde{K} \subset \overline{V_\epsilon(K)}$ then*

$$G_c^3(\widetilde{K}, \epsilon) \subset G_c^3(K, 3\epsilon). \tag{8.34}$$

PROOF. Openness and (8.34) follow as in Lemma 6.6. To prove density let $f \in H(X)$ and $\epsilon_1 > 0$ be given. By shrinking we can assume $\epsilon_1 < \epsilon/2$ and that $d(f, g) < \epsilon_1$ implies condition (6.13) with $\delta = \epsilon/2$.

Since X is spongy and generalized homogeneous, Lemma 6.6 implies that $G^3(K, \epsilon_1/2, 1, \mathcal{U})$ is dense with $\mathcal{U}$ the basis of all clopen subset of X. Choose $f_1 \in H(X)$ with $d(f, f_1) < \epsilon_1/2$ and $f_1 \in G^3(K, \epsilon_1/2, 1, \mathcal{U})$.

There exists $\widetilde{U}$ clopen and inward for f_1 with $(\epsilon_1/2, k)$-decomposition $\{\widetilde{U}_i\}$ and such that $d(K, \widetilde{U}) < \epsilon_1/2 + d(K, CT(f))$. Hence, $f_1^k(\widetilde{U}_0) \subset\subset \widetilde{U}_0$. If equality holds then $\widetilde{U}$ is f_1 invariant and so $f_1 \in G_c^3(K, \epsilon)$. Otherwise there exists U_0 clopen with $f_1^k(\widetilde{U}_0) \subset U_0 \subset \widetilde{U}_0$ and $\widetilde{U}_0 \backslash U_0 \neq \emptyset$. Define $h \in H(X)$ to be 1_X on $X \backslash \widetilde{U}_0$ and such that

$$hf^k(U_0) = U_0 \text{ and } h(\widetilde{U}_0 \backslash f^k(U_0)) = \widetilde{U}_0 \backslash U_0. \tag{8.35}$$

Since the diameter of $\widetilde{U}_0$ is less than $\epsilon_1/2$, $d(g, f_1) < \epsilon_1/2$ where $g = h \circ f_1$.

With $U_i = f_1^i(U_0) = g^i(U_0)$ for $i = 0, \dots, k-1$, we see that $\{U_i\}$ is an $(\epsilon_1/2, k)$-periodic decomposition for the clopen g-invariant set $U = \bigcup_{i-0}^{k-1} U_i$. As in Lemma 2.1, $U \cap \widetilde{U}_i \neq \emptyset$ for $i = 0, \dots, k-1$ implies $d(\widetilde{U}, U) < \epsilon_1/2$. Hence, $d(K, U) < \epsilon_1 + d(K, CT(f)) \leq \epsilon + d(K, CT(g))$ by (6.13).

Thus, $d(g, f) < \epsilon_1$ and $g \in G_c^3(K, \epsilon)$. □

For X a Cantor set and $\mathcal{U}$ the basis of clopen subsets of X define:

(8.36) $\mathcal{H}_{3,c}$ is the set of $f \in H$ satisfying (8.36.1):

(8.36.1) : For all nonempty closed subsets K, K_1 of X, all $\epsilon > 0$ and all integers $m > 0$, $f, f^{-1} \in G^1(K, K_1, \epsilon, \mathcal{U}) \cap G_c^3(K, \epsilon)$, and $\mathcal{P}_m(f) = \mathcal{C}_m(f)\}$.

THEOREM 8.10. *For a Cantor set X, $\mathcal{H}_{3,c}$ is a dense, G_δ conjugacy invariant subset of $H(X)$, with*

$$\mathcal{H}_{3,c} \subset \mathcal{H}_1. \tag{8.37}$$

If $f \in \mathcal{H}_{3,c}$ then $\Omega(f) = \mathcal{C}(f)$ and the set of $\Pi_{\infty,c}(f)$ chain components is a dense G_δ subset of $CT(f)$ and a dense G_δ subset of $\mathcal{B}_f$. Furthermore, $\Pi_{\infty,c}(f)$ is a G_δ set dense in $\Omega(f) = \mathcal{C}(f)$.

PROOF. For the first result proceed just as in Theorem 6.7 using Lemma 8.9. Prove that the $\Pi_{\infty,c}(f)$ sets are G_δ in $CT(f)$ and $\mathcal{B}_f$, and that $\Pi_{\infty,c}(f)$ is G_δ in X as in Lemma 2.5. The density results follow as in Theorem 6.8. □

For homeomorphisms in $\mathcal{H}_6 \cap \mathcal{H}_{3,c}$ we can provide some additional description.

A continuous map $\pi : X_1 \to X_2$ is called an *almost homeomorphism* when it is surjective and the G_δ set

$$\text{Inj}(\pi) \equiv \{x \in X_1 : \pi^{-1}(\pi(x)) = \{x\}\} \tag{8.38}$$

is dense in X_1.

THEOREM 8.11. *For a Cantor set X assume that $f \in \mathcal{H}_6 \cap \mathcal{H}_{3,c}$. There exist Cantor sets X_1, X_0, a homeomorphism $f_1 \in H(X_1)$ and continuous surjective maps $\pi_1 : X \to X_1$, $\pi_0 : X_1 \to X_0$ which satisfy the following conditions:*

(a) *The surjection π_1 maps f to f_1 and π_0 maps f_1 to 1_{X_0}. That is,*

$$\pi_1 \circ f = f_1 \circ \pi_1 \text{ and } \pi_0 \circ f_1 = \pi_0. \tag{8.39}$$

(b) *The homeomorphism f_1 on X_1, is equicontinuous and restricts to an adding machine on each fiber of π_0, i.e. on each closed invariant set $\pi_0^{-1}(z)$ for $z \in X_0$.*

(c) *π_1 is injective at each point of $\Pi_{\infty,c}(f)$. That is,*

$$\Pi_{\infty,c}(f) \subset \text{Inj}(\pi_1). \tag{8.40}$$

In particular, the restriction of π_1 to $\Omega(f) = \mathcal{C}(f)$ is an almost homeomorphism onto X_1.

PROOF. To say that f is in $\mathcal{H}_6$ is to say that f represents every surjective relation on $[n]$ for $n = 1, 2, \dots$. In particular, f represents every surjective map on $[n]$. For a finite set a surjective map is a permutation and so is a union of disjoint cycles. So f represents every permutation.

It will be easier to use partitions than numbered partitions here. If $\mathcal{A}$ is a partition of X, it is easy to see that

$$f\mathcal{A} = \mathcal{A} \Leftrightarrow f \text{ permutes the elements of } \mathcal{A}. \tag{8.41}$$

When (8.41) holds we will call $\mathcal{A}$ an *invariant partition.* We will call $\mathcal{A}$ a *fixed partition* when it satisfies the stronger condition:

$$f(A) = A \text{ for all } A \in \mathcal{A}, \text{ i.e., } f^{\mathcal{A}} = 1_{\mathcal{A}}. \tag{8.42}$$

Let us write $\mathcal{A}_1 \to \mathcal{A}_2$ for partitions $\mathcal{A}_1$ and $\mathcal{A}_2$ when $\mathcal{A}_1$ *refines* $\mathcal{A}_2$, i.e. each element of $\mathcal{A}_1$ is included in a unique member of $\mathcal{A}_2$. These inclusions induce a categorical surjection from $\mathcal{A}_1$ to $\mathcal{A}_2$ (which is why we use the mapping symbol $\to$ for the ordering). That is, if $\mathcal{A}_1 \to \mathcal{A}_2$ and $\mathcal{A}_2 \to \mathcal{A}_3$ then $\mathcal{A}_1 \to \mathcal{A}_3$ with the composed surjection, and the map $\mathcal{A} \to \mathcal{A}$ is the identity. Since the join $\mathcal{A}_1 \wedge \mathcal{A}_2$ refines $\mathcal{A}_1$ and $\mathcal{A}_2$, the family of partitions of X is directed by this ordering. Because the join of two invariant (or fixed) partitions is invariant (resp. fixed) these families of partitions are directed as well. Furthermore, if $\mathcal{A}_1$ and $\mathcal{A}_2$ are invariant partitions with $\mathcal{A}_1 \to \mathcal{A}_2$ then the categorical surjection maps $f^{\mathcal{A}_1}$ to $f^{\mathcal{A}_2}$.

Define X_1 to be the inverse limit of the family of invariant partitions. Define f_1 on X_1 to be the inverse limit of the permutations $f^{\mathcal{A}}$ on $\mathcal{A}$, for all invariant partitions $\mathcal{A}$. Because it is the inverse limit of a countable family of finite sets, X_1 is compact, metrizable and zero-dimensional. Because f_1 is the inverse limit of permutations every point of X_1 is an equicontinuity point and so f_1 is equicontinuous.

For $x \in X$ define $\pi^{\mathcal{A}}(x)$ to be the element of the partition $\mathcal{A}$ which contains x, i.e. the $\sim_{\mathcal{A}}$ equivalence class of x. If $\mathcal{A}_1 \to \mathcal{A}_2$ then $\pi^{\mathcal{A}_1}$ composed with the surjection is $\pi^{\mathcal{A}_2}$. So for $\mathcal{A}$ invariant, the $\pi^{\mathcal{A}}$'s induce a continuous surjection $\pi_1 : X \to X_1$ which clearly maps f to f_1. Furthermore, if $x, y \in X$, then

$$\pi_1(x) = \pi_1(y) \Leftrightarrow x \sim_{\mathcal{A}} y \text{ for every invariant partition } \mathcal{A}. \tag{8.43}$$

In particular, $x \in \text{Inj}(\pi_1)$ iff for every $\epsilon > 0$ there is an invariant partition $\mathcal{A}$ such that $\pi^{\mathcal{A}}(x)$ has diameter less than ϵ. If $x \in \Pi_{\infty,c}(f)$ then $x \in U_0$ where $\{U_i\}$ is an (ϵ, k)-partition for a clopen f-invariant set U. Then $\mathcal{A} = \{X \backslash U, U_0, \dots, U_{k-1}\}$ is an invariant partition with $\pi^{\mathcal{A}}(x) = U_0$. This proves (8.40). If $f \in \mathcal{H}_{3,c}$ then by Theorem 8.10, $\Pi_{\infty,c}(f)$ is dense in $\Omega(f) = \mathcal{C}(f)$ and so f is an almost homeomorphism. This proves part (c).

Similarly, let X_0 be the inverse limit of the fixed partitions. The inclusion of the set of fixed partitions into the set of invariant partitions induces the map $\pi_0 : X_1 \to X_0$. Since $f^{\mathcal{A}} = 1_{\mathcal{A}}$ for $\mathcal{A}$ fixed, the map induced on X_0 is the identity.

For each invariant partition $\mathcal{A}$ there is a largest fixed partition, which we will denote $[\mathcal{A}]$, of which $\mathcal{A}$ is a refinement. Since $f^{\mathcal{A}}$ is a permutation, the set $\mathcal{A}$ decomposes into a finite list of disjoint cycles, i.e. the orbits of $f^{\mathcal{A}}$. $[\mathcal{A}]$ is obtained by taking the union of each cycle. Of course, if $\mathcal{A}$ is fixed already then $[\mathcal{A}] = \mathcal{A}$.

By definition of the inverse limit, a point w of X_1 consists of a choice $w_{\mathcal{A}} \in \mathcal{A}$ for every invariant partition $\mathcal{A}$, with the choices coherent with respect to the surjections $\mathcal{A}_1 \to \mathcal{A}_2$. Similarly, for $z \in X_0$ $z_{\mathcal{A}} \in \mathcal{A}$ for every fixed partition $\mathcal{A}$. The map $\pi_0 : X_1 \to X_0$ satisfies for $w \in X_1$ and $\mathcal{A}$ any fixed partition:

$$\pi_0(w)_{\mathcal{A}} = w_{\mathcal{A}}. \tag{8.44}$$

Hence, for $w_1, w_2 \in X_1$:

$$\pi_0(w_1) = \pi_0(w_2) \Leftrightarrow w_{1[\mathcal{A}]} = w_{2[\mathcal{A}]} \tag{8.45}$$

for every invariant partition $\mathcal{A}$. So $w_{1\mathcal{A}}$ and $w_{2\mathcal{A}}$ lie in the same cycle of $f^{\mathcal{A}}$.

It follows that on each fiber $\pi_0^{-1}(z)$, f_1 restricts to an inverse limit of cycles. So f_1 on $\pi_0^{-1}(z)$ is either a periodic orbit or an adding machine.

Because $f \in \mathcal{H}_6$ every permutation can be realized. For an arbitrary positive integer n, let $\mathcal{A}$ be a partition so that $f^{\mathcal{A}}$ is a single cycle of length n. Then $\mathcal{A}$ is an invariant partition and the projection from X_1 to $\mathcal{A}$ by $w \mapsto w_{\mathcal{A}}$ maps f_1 onto the cycle $f^{\mathcal{A}}$. As each fiber $\pi_0^{-1}(z)$ is invariant, each fiber maps onto $\mathcal{A}$. As n was arbitrary it follows that each $\pi_0^{-1}(z)$ is infinite and so f_1 is an adding machine on each $\pi_0^{-1}(z)$. This implies X_1 is perfect and so is a Cantor set.

It remains to prove that X_0 is perfect. Assume that $z \in X_0$. It suffices to show that if $\mathcal{A}$ is an arbitrary fixed partition then there exists $z_1 \in X_0$ such that $z_{\mathcal{A}} = z_{1\mathcal{A}}$ but $z \neq z_1$.

Let $z_{\mathcal{A}} = A \in \mathcal{A}$. A is a clopen invariant set so it meets, and hence contains, some chain component B for f. By Theorem 8.10 we can choose B to lie in $\Pi_{\infty,c}$. Since $f \in \mathcal{H}_{3,c} \subset \mathcal{H}_1$, Theorem 6.14(d) implies that B is nowhere dense in X. So there exists a clopen invariant set U with $B \subset U$ but close enough to B that $A \backslash U \neq \emptyset$. Let $\mathcal{A}_1 = \mathcal{A} \wedge \{U, X\backslash U\}$. This is a fixed partition and $z_{\mathcal{A}_1}$ is either $A \cap U$ or $A \cap (X \backslash U)$. Choose $z_1 \in X_0$ so that $z_{\mathcal{A}_1}$ is the other choice. Then $z \neq z_1$, $z_{1\mathcal{A}} = A = z_{\mathcal{A}}$ as required. □

We in fact proved that on each fiber, f_1 projects onto a cycle length k for every positive integer k. Such an adding machine is called a *universal adding machine* . If $\{k_n\}$ is the sequence of positive integers with $k_n | k_{n+1}$ and $k_n \to \infty$ which was used to define the adding machine construction, then the adding machine is universal if for every positive integer k, $k | k_n$ for some n. This says exactly that the sequence $\{k_n\}$ is cofinal in the directed set $\mathbf{Z}^*$ consisting of positive integers ordered by divisibility. So up to isomorphism a universal adding machine is the inverse limit of the sequence of cyclic groups $\{\mathbf{Z}_k\}$ indexed by $k \in \mathbf{Z}^*$. So we will speak of *the* universal adding machine.

COROLLARY 8.12. *For a Cantor set X, if $f \in \mathcal{H}_6$, then every chain component for f has the universal adding machine as a factor.*

PROOF. $f \in \mathcal{H}_{3,c}$ was used only to prove that X_0 was perfect and that π_1 was an almost homeomorphism on $\mathcal{C}(f)$. In any case, $\pi_0\pi_1$ maps any chain component B of f onto a chain component for 1_{X_0} which is a singleton since X_0 is compact and zero-dimensional. Hence, $\pi_1(B)$ lies in some fiber $\pi_0^{-1}(z)$. Since f_1 is minimal on the fiber, $\pi_1(B) = \pi_0^{-1}(z)$. Hence, π_1 maps f on B to the universal adding machine f_1 on $\pi_0^{-1}(z)$. □

We conclude by observing that neither set $\mathcal{H}_6$ nor $\mathcal{H}_{3,c}$ is included in the other. Begin with $f \in \mathcal{H}_6$ and let $h \in H(X)$ be any minimal homeomorphism which is not an adding machine. For example, in Chapter 1 of Auslander (1988) the Morse minimal homeomorphism is described. It is an infinite subshift and so is not equicontinuous because it is expansive. Use Corollary 8.7 to construct $f_1 = s^{-1}(f \times h)s \in \mathcal{H}_6$. The restriction of f_1 to any nonempty closed invariant subset of X has h as a factor and so is not equicontinuous. Hence, $\Pi_\infty(f_1) \cup \Pi_\infty(f_1^{-1}) = \emptyset$. So by Theorem 6.4, $f_1 \notin \mathcal{H}_1$.

On the other hand, begin with $f \in \mathcal{H}_{3,c}$ and choose B a $\Pi_{\infty,c}(f)$ chain component. On the quotient space X_1 obtained by smashing B to a point p, f induces the homeomorphism f_1 with fixed point p. Because B is dynamically isolated it is not hard to show that as a homeomorphism of X_1, $f_1 \in \mathcal{H}_{3,c}$. However, since f_1 has a fixed point, Corollary 8.5 implies $f_1 \notin \mathcal{H}_6$.

CHAPTER 9

The Circle

Overview: Here we examine what our results give when applied to the circle. Because the circle is the simplest closed manifold, some of the consequences are particularly simple, and others do not apply at all. We are able to give a fairly complete description of the dynamics of a generic homeomorphism of the circle.

Throughout this section X will be the circle S^1 which we will regard as the group $\mathbf{R}/\mathbf{Z}$.

We use the ordering on $\mathbf{R}$ to orient the circle. If a, b are distinct points of S^1 we denote by $[a, b]$ the closed interval consisting of points obtained by moving in the positive direction along the circle from a to b. Hence, $[a, b]$ and $[b, a]$ are closed intervals intersecting at the endpoints and with union the entire circle. As usual, we let (a, b) denote the open interval $[a, b] \setminus \{a, b\}$.

Since the projection map $\pi : \mathbf{R} \to \mathbf{R}/\mathbf{Z}$ is the universal covering space, any homeomorphism f of $\mathbf{R}/\mathbf{Z} = S^1$ lifts to a homeomorphism F of $\mathbf{R}$. Different lifts differ by an integer constant. f is called *orientation preserving* when F is order preserving, i.e. F is a strictly increasing function, while f is called *orientation reversing* when F is order reversing, i.e. decreasing. So if f is orientation preserving then

$$f([a, b]) = [f(a), f(b)] \tag{9.1}$$

while if f is orientation reversing then

$$f([a, b]) = [f(b), f(a)]. \tag{9.2}$$

The order preserving homeomorphisms of $\mathbf{R}$ form a subgroup of index two in the group of homeomorphisms of $\mathbf{R}$. It follows that $H^+(S^1)$, the set of orientation preserving homeomorphisms of the circle, is a clopen subgroup of index two in $H(S^1)$.

9.1. Background

Before describing the generic homeomorphism of the circle, we first review what the rotation number tells us about homeomorphisms of the circle in general. The elementary results which we describe below can be found in Katok and Hasselblatt (1995) Sections 11.1 and 11.2 together with Nitecki (1971) Section 1.1.

For f orientation preserving with lift F on $\mathbf{R}$ we define the *rotation number*

$$\tau(F) = \lim_{n \to \infty} \frac{F^n(t)}{n}. \tag{9.3}$$

As the labelling suggests the limit is independent of the choice of t in $\mathbf{R}$. The lift of $f \in H^+(S^1)$, F satisfies

$$F(x + 1) = F(x) + 1. \tag{9.4}$$

Hence, adding an integer to F adds the same integer to $\tau(F)$. So the rotation number $\tau(f)$ is well defined as the congruence class of $\tau(F)$ mod $\mathbf{Z}$.

For $f \in H^+(S^1)$ the rotation number (mod $\mathbf{Z}$) is a conjugacy invariant with respect to orientation preserving changes of variables. That is,

$$\tau(hfh^{-1}) = \tau(f) \quad f, h \in H^+(S^1). \tag{9.5}$$

If the conjugating homeomorphism is orientation reversing then the rotation number is reversed in sign:

$$\tau(hfh^{-1}) = -\tau(f) \quad f \in H^+(S^1),\ h \in H(S^1)\backslash H^+(S^1). \tag{9.6}$$

The fundamental example is a *rotation* on the circle which lifts to the *translation* map F_c defined by $F_c(t) = t + c$. Clearly,

$$\tau(F_c) = c. \tag{9.7}$$

The associated circle map f_c is minimal when c is irrational and in this case $\mathrm{Per}(f_c) = \emptyset$. When c is rational, with $kc \in \mathbf{Z}$ then $(f_c)^k = 1_{S^1}$ and so the entire circle consists of periodic points.

In general, for $f \in H^+(S^1)$, $\tau(f)$ is rational iff f has some periodic points. We will distinguish three cases where periodic points occur.

Case 1: $f \in H^+(S^1)$ and $\tau(f) = 0$ mod $\mathbf{Z}$. In this case every periodic point is fixed. By rotating to change coordinates we can assume that 0 is a fixed point and then choose the lift F so that $F(0) = 0$. Then by (9.4) F restricts to an increasing function on $[0, 1] \subset \mathbf{R}$ which fixes the endpoints. Furthermore, the other fixed points of f lift to fixed points of F.

In any case, there is a unique lift F of f such that the fixed points of f lift to fixed points of F. We will call this *the fixed point lift* for f.

Case 2: $f \in H^+(S^1)$ and $\tau(f) = p/k$ mod $\mathbf{Z}$ with p, k integers such that $0 < p < k$ and p is relatively prime to k. In this case, every periodic point x has period k. Furthermore, for each periodic point x the order of the list of k distinct points $\{x, f(x), \dots, f^{k-1}(x)\}$ on the circle agrees with the order of the list $\{0, p/k, 2p/k, \dots, (k-1)p/k\}$ on the circle. Since p is relatively prime to k it is a generator for the cyclic group $\mathbf{Z}_k$. This reordering is given by the permutation σ of the set $\{0, \dots, k-1\}$ uniquely defined by

$$\sigma(i) \equiv ip \mod k. \tag{9.8}$$

So, in particular, for every periodic point x the first positive integer r such that the open interval $(x, f^r(x))$ contains no members of the orbit of x is the unique $r \in \{0, \dots, k-1\}$ such that

$$1 \equiv rp \bmod k. \tag{9.9}$$

Thus, this number r is the same for every $x \in \mathrm{Per}(f)$ and is obtainable via (9.9) from the rotation number.

Of course, the fixed point Case 1 is a special case of the periodic Case 2 (we let $p = k = 1$ in that case). We distinguish them because we study the periodic case by first analyzing the increasing function f^k on the interval $[x, f^r(x)]$ following the fixed point case, and then by moving the results to $[f^r(x), f^{2r}(x)], [f^{2r}(x), f^{3r}(x)], \dots$ by using the homeomorphism f^r as a conjugacy.

Since f^k is an orientation preserving homeomorphism with fixed points it has a unique fixed point lift. However, when $k > 1$ this lift is never of the form F^k for a lift F of f. To see this notice that an order preserving homeomorphism of $\mathbf{R}$ has no periodic points other than fixed points.

Case 3: $f \in H(S^1)\backslash H^+(S^1)$. When f is orientation reversing there are exactly two distinct fixed points x, y of f. By (9.2), $f([x,y]) = [y,x]$. Since $f^2 \in H^+(S^1)$ with fixed points, $\mathrm{Per}(f) = \mathrm{Fix}(f^2)$. Hence, any periodic points other than $\{x,y\}$ have period two. Again we reduce to the fixed point Case 1 by first analyzing f^2 on $[x,y]$ and then by using f as a conjugacy moving the results to $[y,x]$.

If 0 is one of the fixed points, x then we can choose a lift F of f so that $F(0) = 1$. Because F is the lift of an orientation reversing homeomorphism $F(1) = 0$. The restriction of F to $[0,1]$ has a unique fixed point, the lift of y, and F^2 is the fixed point lift for f^2.

In each case the set of periodic points is closed with $\mathrm{Per}(f) = \mathrm{Fix}(f^k)$ for a unique smallest positive integer k, e.g. $k = 1$ in Case 1, and $k = 2$ in Case 3, and in each case $f^k \in H^+(S^1)$ and has fixed points. The open set $S^1\backslash\mathrm{Per}(f)$ is the union of a countable family of pairwise disjoint open intervals which we will call the *complementary intervals* for f. When $\mathrm{Per}(f)$ is not a singleton each complementary interval can be written uniquely as (a_0, a_1) with a_0, a_1 two distinct elements of $\mathrm{Per}(f)$. By (9.1)

$$f^k([a_0, a_1]) = [a_0, a_1] \tag{9.10}$$

with $f^k(a_0) = a_0$, $f^k(a_1) = a_1$ and $f^k(x) \neq x$ for $x \in (a_0, a_1)$. We will call the interval (a_0, a_1) an *up interval* if for all $x \in (a_0, a_1)$

$$[x, f^k(x)] \subset (a_0, a_1). \tag{9.11}$$

This is equivalent to saying that for F the fixed point lift of f^k, $F(t) > t$ for every t in the intervals of $\mathbf{R}$ which are the lifts of (a_0, a_1). Similarly, we call (a_0, a_1) a *down interval* if $F(t) < t$ on the lifts, or equivalently if for all $x \in (a_0, a_1)$.

$$[f^k(x), x] \subset (a_0, a_1). \tag{9.12}$$

Again excluding the case where $\mathrm{Per}(f)$ is a singleton, we will call $\{a_0, a_1\}$ an *endpoint pair* for f, and an *up endpoint pair* or *down endpoint pair* corresponding to the nature of the interval (a_0, a_1).

For x in an up interval (a_0, a_1) $\{[f^j(x), f^{j+1}(x)] : j \in \mathbf{Z}\}$ is a bi-infinite sequence of closed intervals with union (a_0, a_1) and with the only intersections at the common endpoint of successive intervals. Furthermore, as $j \to +\infty$ the intervals converge to $\{a_1\}$ and as $j \to -\infty$ the intervals converge to $\{a_0\}$.

Similarly, for x in a down interval (a_0, a_1) the intervals in the sequence $\{[f^{j+1}(x), f^j(x)] : j \in \mathbf{Z}\}$ converge to $\{a_0\}$ as $j \to +\infty$ and to $\{a_1\}$ as $j \to -\infty$.

9.2. The class $\mathcal{H}_1$ on S^1

Now we consider the generic homeomorphisms on S^1. First note that the closed intervals are sponges in S^1. So S^1 is spongy and Theorem 6.2 applies to the evaluation action of $H(S^1)$ on S^1. We let $\mathcal{U}$ be the basis of all open subsets of S^1 and so omit it in the label for the dense G_δ subset H^*_{12}. On the other hand, the connected open subsets of S^1 are, except for S^1 itself, the open subintervals of S^1, whose closures – other than S^1 – are the one dimensional topological balls in S^1.

So with $\mathcal{U}_B$ the base of balls defined in (2.21) we can apply Proposition 2.7(c) to get for any $f \in H(S^1)$

$$\Pi_\infty(f) = \Pi_\infty(f, \mathcal{U}_B). \tag{9.13}$$

THEOREM 9.1. *Let $\mathcal{H}_1$ be the subset of $H(S^1)$ defined by (6.5) with $\mathcal{U}$ the basis of all open subsets of S^1.*

(a) *$\mathcal{H}_1$ is a dense, G_δ, conjugacy invariant subset of $H(S^1)$.*

(b) *A homeomorphism f on S^1 satisfies $f \in \mathcal{H}_1$ iff $\mathrm{Per}(f)$ is a Cantor set in S^1 and for every open subset O of S^1, the following condition* (*) is satisfied:

(*) if O contains a periodic point, then O contains both a complementary up interval and a complementary down interval.

(c) Rotation number and orientation are the only obstructions to topological conjugacy for maps in $\mathcal{H}_1$. Specifically, for $f_1, f_2 \in \mathcal{H}_1$.

(i) If either $f_1, f_2 \in H(S^1)\backslash H^+(S^1)$ or $f_1, f_2 \in H^+(S^1)$ with $\tau(f_1) = \tau(f_2)$ then there exists $h \in H^+(S^1)$ such that

$$f_1 = h f_2 h^{-1}. \tag{9.14}$$

(ii) If $f_1, f_2 \in H^+(S^1)$ with $\tau(f_1) = -\tau(f_2)$ then there exists $h \in H(S^1)\backslash H^+(S^1)$ such that (9.14) holds.

PROOF. Part (a) is Theorem 6.2 in this case.

Now assume that $f \in \mathcal{H}_1$. By Theorem 6.4(a) $\Omega(f)$ is a Cantor subset of S^1 and $\Pi_\infty(f)$ is a dense subset of $\Omega(f)$. By (9.13) and (2.24), $\Pi_\infty(f)$ is contained in the closure of $\mathrm{Per}(f)$. Hence, $\mathrm{Per}(f) \neq \emptyset$ and so f is one of our three cases above. In particular, $\mathrm{Per}(f) = \mathrm{Fix}(f^k)$ for suitable k, and so is closed. Hence, we have

$$\Omega(f) = \overline{\Pi_\infty(f)} = \mathrm{Per}(f) = \mathrm{Fix}(f^k). \tag{9.15}$$

Suppose some chain component B for f meets $S^1\backslash\mathrm{Per}(f)$ and so contains a point x in some complementary interval (a_0, a_1) for f. For every x_1 in the interval, $\omega(x_1, f^k) \cup \alpha(x_1, f^k) = \{a_0, a_1\}$. Since B is closed and f-invariant this implies with $x_1 = x$ that $a_0, a_1 \in B$. Then applied to an arbitrary x_1 we see that $x_1 \in \mathcal{C}(B, f) \cap \mathcal{C}(B, f^{-1})$ which is B. Hence, B contains the entire open interval (a_0, a_1). However, Theorem 6.4 (d) says that $\mathcal{C}(f)$ is nowhere dense in S^1. This contradiction implies that $\mathcal{C}(f)$ is contained in $\mathrm{Per}(f)$ and so we have

$$\mathcal{C}(f) = \mathrm{Per}(f) = \Omega(f). \tag{9.16}$$

If B is a chain component for f then by Proposition 1.1, $B = \bigcup_{i=0}^{k-1} f^i(B_0)$ where B_0 is a chain component for f^k. But f^k is the identity on $\mathcal{C}(f)$ and every chain component is chain transitive. As mentioned in Section 1 the chain transitive subsets for the identity on X are exactly the connected subsets of X. Since B_0 is contained in the Cantor set $\Omega(f)$ it follows that B_0 is a singleton and so B is the associated periodic orbit.

Now assume that O is open and $O \cap \mathcal{C}(f) \neq \emptyset$. Since $\Pi_\infty(f)$ is dense in $\mathcal{C}(f)$ and $\mathcal{C}(f)$ is perfect, $O \cap \Pi_\infty(f)$ is an infinite set, contained in $\mathrm{Per}(f)$. In Case 3 we can avoid the two fixed points, and so we can find $x \in O \cap \Pi_\infty(f)$ with x a periodic point of period k. Choose $\epsilon > 0$ with

$$\epsilon < d(f^i(x), f^j(x)) \text{ when } 0 \leq i < j < k \tag{9.17}$$

and with $V_\epsilon(x) \subset O$. There is an (ϵ, k_1)-periodic decomposition $\{U_i\}$ of an inward set U for f with $x \in U_0$. Since $f^i(x) \in U_i$ and the U_i's are disjoint for $0 \leq i < k_1$, we must have $k_1 \leq k$. On the other hand, $f^{k_1}(x)$, $x \in U_0$ which has diameter less than ϵ. So $f^{k_1}(x) = x$ which implies $k_1 = k$. By (9.13) we can assume that the U_i's are subintervals of S^1. In particular, $U_0 = [x_0, x_1]$. Furthermore $[f^k(x_0), f^k(x_1)] \subset (x_0, x_1)$. As neither x_0 nor x_1 are fixed by f^k, each is contained in a complementary interval for f^k. The open interval $(x_0, f^k(x_0))$ is contained in $U_0 \backslash f^k(U_0)$ and so is disjoint from $\mathcal{C}(f)$. Hence, it is contained entirely in the complementary interval containing x_0. So the complementary interval containing x_0 is an up interval and similarly the interval containing x_1 is a down interval.

By Theorem 6.3, the attractor $A = \omega(U, f)$ contains a repellor R_1 and by (6.10) the basin $W^u_f(R_1) \subset A$. Hence, $R_1 \subset A^\circ$. By Lemma 2.1 $R_1 \cap U_i : i \in \mathbf{Z}$, is a periodic decomposition for R_1. Because R_1 is an invariant set it meets $\Omega(f)$ and so by periodicity $R_1 \cap U_0$ meets $\Omega(f)$. So $(A \cap U_0)^\circ$ meets $\Omega(f) = \mathrm{Per}(f)$.

We can repeat the previous argument with O replaced by $(A \cap U_0)^\circ$ and obtain a closed interval $[y_0, y_1] \subset (A \cap U_0)^\circ$ which is inward for f^k. Again y_0 is contained in some complementary up interval (a_0, b_0) so that $\{a_0\} = \alpha(y_0, f^k)$ and $\{b_0\} = \omega(y_0, f^k)$. But now $A \cap U_0$ is closed and f^k invariant so that $a_0, b_0 \in A \cap U_0$. Furthermore, $A \cap U_0 = \bigcap_{i=0}^{\infty} f^{ik}([x_0, x_1])$, and so $A \cap U_0$ is connected. As $a_0, b_0, y_0 \in A \cap U_0$ it follows that the entire up interval $[a_0, b_0] \subset A \cap U_0 \subset O$. Similarly the down interval containing y_1 is contained in O.

Next we show that $\mathcal{H}_1$ meets $H(S^1) \backslash H^+(S^1)$ and $\tau^{-1}(r) = \{f \in H^+(S^1) : \tau(f) = r \bmod \mathbf{Z}\}$ for every rational number r. Since $\mathcal{H}_1$ is dense in $H(S^1)$ it suffices to show that each of these sets has a nonempty interior.

$H^+(S^1)$ is a closed subgroup. So $H(S^1) \backslash H^+(S^1)$ is open.

We can suppose that $g \in H^+(S^1)$ $\tau(g) = p/k \bmod \mathbf{Z}$ for positive integers p, k with $0 < p \leq k$. Let $x \in \mathrm{Per}(g)$ so that x has period k and p is determined by the ordering of the list $\{x, g(x), \ldots, g^{k-1}(x)\}$ on the circle. Assume ϵ satisfies the analogue of (9.17) for g. Following Proposition 4.3 we can perturb g to obtain g_1 and an $(\epsilon/2, k)$-periodic decomposition $\{U_0, \ldots U_{k-1}\}$ of a g_1 inward set U so that $g^i(x) \in U_i$ for $i = 0, \ldots, k-1$ and each U_i is a closed interval. For any g_2 sufficiently close to g_1, U is still inward for g_2 with periodic decomposition $\{U_0, \ldots, U_{k-1}\}$. On the closed interval U_0, g_2^k has some fixed point y and the ordering of the points $\{y, g_2(y), \ldots, g_2^{k-1}(y)\}$ agrees with the ordering of $\{U_0, \ldots U_{k-1}\}$ and so with $\{x, g(x), \ldots, g^{k-1}(x)\}$. Hence, $\tau(g) = \tau(g_2)$ for all g_2 close enough to g_1.

Now suppose that for $\alpha = 1, 2$ $f_\alpha \in H(S^1)$ has $\mathrm{Per}(f_\alpha)$ a Cantor set in S^1 and satisfies condition (*) from Theorem 9.1(b). We will show that if either $f_1, f_2 \in H(S^1) \backslash H^+(S^1)$ or $f_1, f_2 \in H^+(S^1)$ with $\tau(f_1) = \tau(f_2)$ then f_1 is conjugate to f_2 by an element of $H^+(S^1)$. In particular, since $\mathcal{H}_1$ is conjugacy invariant it will follow that $f_2 \in \mathcal{H}_1$ if f_1 is. We have just seen that $H(S^1) \backslash H^+(S^1)$ and every rational rotation number has some $\mathcal{H}_1$ element. Thus the proof of (c) will then be complete. We begin by showing how to reduce (c) (ii) to (c) (i).

If $f_1, f_2 \in H^+(S^1)$ and $\tau(f_1) = -\tau(f_2)$ then choose h_1 an arbitrary element of $H(S^1) \backslash H^+(S^1)$, e.g. $h_1(x) = -x$ in $\mathbf{R}/\mathbf{Z}$, and let $f_3 = h_1 f_1 h_1^{-1}$. Clearly, f_3 satisfies the conditions of (c), $f_3 \in H^+(S^1)$ and by (9.6), $\tau(f_3) = -\tau(f_1) = \tau(f_2)$. Applying (c) (i) to f_3 we obtain $h_2 \in H^+(S^1)$ such that $h_2 h_1 f_1 h_1^{-1} h_2^{-1} = f_2$. Thus, $h = h_2 h_1$ is an element of $H(S^1) \backslash H^+(S^1)$ providing a conjugacy from f_1 to f_2.

To prove our remaining conjugacy result we consider our three cases.

Case 1: $\mathrm{Fix}(f_1)$ and $\mathrm{Fix}(f_2)$ are Cantor sets. By conjugating by rotations we can assume that for both maps 0 is a fixed point and is not one of the countably many endpoints of complementary intervals. Let F_1 and F_2 be the fixed point lifts of f_1 and f_2 respectively. So each is an order preserving homeomorphism of $I = [0,1]$ with $\mathrm{Fix}(F_1)$ and $\mathrm{Fix}(F_2)$ Cantor subsets of I each containing 0 and 1.

Recall the proof that two such Cantor sets C_1 and C_2 are homeomorphic via an order preserving homeomorphism of I. The first step is the classic result that if D_1 and D_2 are two countable totally ordered sets with a maximum and a minimum element and each is order dense, i.e. between any two elements there exists a third, then there is an order preserving bijection between them. For the second step, regard the left endpoint of each endpoint pair and 0 and 1 as a countable ordered set D_1, dense in the Cantor set C_1 and similarly consider D_2 in C_2. Choose an order isomorphism and map the right endpoints so that matching pairs go with matching pairs. The closure of this association defines an order preserving homeomorphism from C_1 to C_2. For the third step the homeomorphism is extended, linearly, across each complementary interval to obtain a homeomorphism of I.

Our proof parallels this one. For the first step, suppose that D_1 is the disjoint union of D_1^+, D_1^- and the maximum and minimum elements of D_1, and between any two elements of D_1 are members of both D_1^+ and D_1^-. Suppose that D_2 is similarly subdivided. It is an easy exercise to adjust the original inductive construction to obtain an order preserving bijection from D_1 to D_2 so that D_1^+ is mapped to D_2^+ and D_1^- is mapped to D_2^-. Now for $\alpha = 1, 2$ let D_α^+ be the left endpoints of up intervals for F_α and D_α^- be the left endpoints of down intervals for F_α. If (a_0, b_0) and (a_1, b_1) are two complementary intervals in I with $b_0 \le a_1$, then $b_0 \ne a_1$ and (b_0, a_1) is not a complementary interval because either possibility would imply that b_0 is an isolated point in $\mathrm{Fix}(F_\alpha)$. Hence, the open interval (b_0, a_1) meets $\mathrm{Fix}(F_\alpha)$ and so by Theorem 9.1b(*) it contains both up and down intervals.

Selecting an order preserving bijection which maps D_1^+ to D_2^+ and D_1^- to D_2^- we close up to obtain a homeomorphism h from $\mathrm{Fix}(F_1)$ onto $\mathrm{Fix}(F_2)$. Now suppose that (a_0, b_0) is a complementary up interval for F_1, so that $(h(a_0), h(b_0))$ is a complementary up interval for F_2. Choose $a_0 < t_1 < b_0$ and $h(a_0) < h(t_1) < h(b_0)$. Extend h to the orbits so that $h(F_1^i(t_1)) = F_2^i(h(t_1))$. Notice that $\{F_1^i(t_1)\}$ and $\{F_2^i(h(t_1))\}$ are bi-infinite increasing sequences. Extend h linearly to map the interval $[t_1, F_1(t_1)]$ to $[h(t_1), F_2(h(t_1))]$. Then on the interval $[F_1^i(t_1), F_1^{i+1}(t_1)]$ define h to be $F_2^i h F_1^{-i}$ mapping to $[F_2^i(h(t_1)), F_2^{i+1}(h(t_1))]$. Proceed similarly on every up and down interval to obtain the required conjugating map $h : I \to I$ such that $hF_1 = F_2 h$. Since $h(0) = 0$ and $h(1) = 1$, this induces an order preserving homeomorphism of the circle mapping f_1 to f_2.

Case 2: $f_1, f_2 \in H^+(S^1)$ and $\tau(f_1) = \tau(f_2) = p/k \bmod \mathbf{Z}$ with $p < k$, p and k relatively prime positive integers. For $\alpha = 1, 2$ select $x_\alpha \in \mathrm{Per}(f_\alpha)$ but not an endpoint of an interval complementary to this Cantor set. The point x_α has period k for f_α and with r defined by (9.9) the interval $(x_\alpha, f_\alpha^r(x_\alpha))$ is disjoint from the f_α orbit of x_α. Because x_α and $f_\alpha^r(x_\alpha)$ are not endpoints, $[x_\alpha, f_\alpha^r(x_\alpha)] \cap \mathrm{Per}(f_\alpha)$ is a Cantor set. Furthermore, by (9.1) f^k is an order preserving homeomorphism on this interval with fixed point set the intersection with $\mathrm{Per}(f_\alpha)$.

We can apply the Case 1 argument to get a homeomorphism

$$h : [x_1, f_1^r(x_1)] \to [x_2, f_2^r(x_2)]$$

which maps f_1^k to f_2^k. For each j, use $f_2^j h f_1^{-j}$, to extend this to a map from the interval $[f_1^j(x_1), f_1^{r+j}(x_1)]$ to $[f_2^j(x_1), f_2^{r+j}(x_2)]$. These extensions fit together to define the required conjugacy because the order of the two orbit sequences

$$\{x_\alpha, f_\alpha(x_\alpha), \dots, f_\alpha^{k-1}(x_\alpha)\}$$

in the circle for $\alpha = 1, 2$ agree because the rotation numbers do. The homeomorphism h is orientation preserving because it is order preserving on the original subinterval.

Case 3: $f_1, f_2 \in H(S^1) \backslash H^+(S^1)$. For $\alpha = 1, 2$ let x_α, y_α be the two fixed points of f_α. If (a_0, b_0) were an interval complementary to $\text{Per}(f_\alpha)$ with endpoint $a_0 = x_\alpha$ then $f_\alpha((x_\alpha, b_0)) = (f_\alpha(b_0), x_\alpha)$, would also be a complementary interval and so x_α would be an isolated point of $\text{Per}(f_\alpha)$. It follows that $[x_\alpha, y_\alpha] \cap \text{Per}(f_\alpha)$ is a Cantor set and is the fixed point set of the order preserving homeomorphism f_α^2 on the interval.

As above we apply the Case 1 argument to get a homeomorphism $h : [x_1, y_1] \to [x_2, y_2]$ mapping f_1^2 to f_2^2. Then extend the homeomorphism to the opposite interval $[y_1, x_1] = h([x_1, y_1])$ by $f_2 h f_1^{-1}$. □

THEOREM 9.2. *Let $\mathcal{H}_1$ be the subset of $H(S^1)$ defined by (6.5).*

(a) *If $f \in \mathcal{H}_1$ then* $\text{Per}(f) = \Omega(f) = \mathcal{C}(f)$ *is a Cantor set in S^1 and the chain components for f are the periodic orbits of f.*

(b) *On $\mathcal{H}_1 \cap H^+(S^1)$ the rotation number function $f \mapsto \tau(f)$, defined from $H^+(S^1)$ to $\mathbf{R}/\mathbf{Z}$, is locally constant. In fact for every rational number r, if we let $\tau^{-1}(r) = \{f \in H^+(S^1) : \tau(f) = r \bmod \mathbf{Z}\}$ then*

$$\tau^{-1}(r) \cap \mathcal{H}_1 \subset (\tau^{-1}(r))^\circ. \tag{9.18}$$

Furthermore, $\tau^{-1}(r) \cap \mathcal{H}_1$ is dense in $\tau^{-1}(r)$.

(c) *For $f \in \mathcal{H}_1$ let $CI(f)$ denote the set of complementary interval end points. Then*

$$\Pi_\infty(f) \cap \Pi_\infty(f^{-1}) = \text{Per}(f) \backslash CI(f). \tag{9.19}$$

PROOF. Results (a) and (b) were proved in passing in the proof of Theorem 9.1. For example, we showed how to approximate any $g \in \tau^{-1}(r)$ by an element of $(\tau^{-1}(r))^\circ$. Density of $\tau^{-1}(r) \cap \mathcal{H}_1$ in $\tau^{-1}(r)$ then follows. On the open set of elements of $H^+(S^1)$ with a given periodic inward set the rotation number is constant.

For simplicity we will look only at the fixed point Case 1 for (c) and leave the extension to the other two cases to the reader. If x is fixed but is not in $CI(f)$ then arbitrarily close to x and on each side there are both up and down complementary intervals. Hence, x is contained in arbitrarily small inward intervals for f and for f^{-1}. Hence, $x \in \Pi_\infty(f) \cap \Pi_\infty(f^{-1})$. For the converse, any $x \in \Pi_\infty(f) \cap \Pi_\infty(f^{-1})$ is clearly periodic, and if it is an endpoint of a complementary interval then it is either not in a small inward interval for f, or it is not in one for f^{-1}. □

9.3. Relative Rohlin property

Because the set $\{\tau(f), -\tau(f)\}$ in $\mathbf{R}/\mathbf{Z}$ is a conjugacy invariant for $f \in H(S^1)$ and because the function $\tau : H^+(S^1) \to \mathbf{R}/\mathbf{Z}$ is surjective and continuous, see e.g. Katok and Hasselblatt (1995), Proposition 11.1.6, it follows that the circle cannot satisfy the Rohlin Property as the Cantor set did. However, modulo the rotation number invariant the results we have obtained are much stronger for the circle.

For each rational $r \in \mathbf{R}/\mathbf{Z}$, $\tau^{-1}(r)$ is a closed conjugacy invariant subset of the Polish group $H^+(S^1)$. $\mathcal{H}_1$ is a dense, G_δ subset of $H(S^1)$ which intersects $\tau^{-1}(r)$ in a single conjugacy class which is dense in $\tau^{-1}(r)$. There are – as we will see – other conjugacy classes in each $\tau^{-1}(r)$ which are dense. But every other conjugacy class in $\tau^{-1}(r)$ is meager, i.e. each is contained in a countable union of closed subsets which are nowhere dense in $\tau^{-1}(r)$. This is obvious because $\tau^{-1}(r)\backslash\mathcal{H}_1$ is meager in $\tau^{-1}(r)$.

For the Cantor set we were able to characterize nicely, in Theorem 8.4, those homeomorphisms whose conjugacy class is dense. It remains open whether for the Cantor set there exists a residual conjugacy class.

To construct other examples of homeomorphisms of the circle with dense conjugacy class we take advantage of the amount of choice in the proof of part (b) in Theorem 9.1. Specifically, in the construction of the order preserving bijection between countable, order dense sets. We will restrict attention to the fixed point Case 1.

Let $f \in \mathcal{H}_1 \cap H^+(S^1)$ with $\text{Fix}(f) \neq \emptyset$. Fix a particular complementary up interval (a_0, b_0) for f. For any $\epsilon > 0$ there exists a conjugating homeomorphism $h \in H^+(S^1)$ which maps f to itself and such that $h([a_0, b_0])$ is a complementary up interval of diameter less than ϵ. Now let $f_1 = f$ on $S^1\backslash(a_0, b_0)$ and on $[a_0, b_0]$ f_1 is an order preserving homeomorphism with a finite number of isolated fixed points in (a_0, b_0). Then since $hfh^{-1} = f$, hf_1h^{-1} agrees with f except on the interval $h([a_0, b_0])$. Consequently,

$$d(hf_1h^{-1}, f) < \epsilon. \tag{9.20}$$

It follows that the closure of the conjugacy class of f_1 contains f. As the closure of a conjugacy class is conjugacy invariant, the closure of the conjugacy class of f_1 contains that of f which is dense in $\tau^{-1}(0)$.

Furthermore, we can sharpen this construction up to get uncountably many distinct conjugacy classes. Let E be an infinite family of up intervals such that the remaining up intervals are still order dense. One can choose $h \in H^+(S^1)$ which maps f to f and each interval of E to an up interval of diameter less than ϵ (there are only finitely many complementary intervals of diameters greater than or equal to ϵ). Let S be an infinite subset of the set of positive integers. Let $\{I_1, I_2, \dots\}$ be an enumeration of the open complementary intervals of E and let $\{n_1, n_2, \dots\}$ list the numbers of S in order. Define f_S so that on I_i, f_S has exactly n_i fixed points. Otherwise, f_S agrees with f. So $\text{Fix}(f_S)$ is the Cantor set $\text{Fix}(f)$ together with a countable set of isolated points. If T were another infinite subset of positive integers and h were a homeomorphism of S^1 mapping f_S to f_T then $h(\text{Fix}(f_S)) = \text{Fix}(f_T)$ and so h maps the perfect part of $\text{Fix}(f_S)$ and $\text{Fix}(f_T)$, namely $\text{Fix}(f)$, to itself. As h would then have to permute the f complementary intervals it would follow that $S = T$. Hence, distinct subsets S yield distinct conjugacy classes. As before $d(hf_Sh^{-1}, f) < \epsilon$ and so, as before, the conjugacy class of each f_S is dense in $H(S^1)$.

By (9.18)

$$\mathcal{H}_1 \cap H^+(S^1) \subset \bigcup_{r\ rational} \left(\tau^{-1}(r)^\circ\right). \tag{9.21}$$

It follows that $\{f \in H^+(S^1) : \tau(f) \text{ is irrational}\}$ is nowhere dense in $H^+(S^1)$. However, for completeness we will consider the irrational case.

We require a smoothing result.

LEMMA 9.3. *If $f \in H^+(S^1)$ and $\epsilon > 0$ then there exists a C^∞ diffeomorphism g in $H^+(S^1)$ such that $\tau(f) = \tau(g)$ and $d(f, g) < \epsilon$.*

PROOF. At this point, this is a fairly straightforward exercise. On the circle, diffeomorphisms are dense in $H^+(S^1)$, and we have seen that for r rational, $\tau^{-1}(r)$ is the closure of its interior, so in this case the result is obvious. In the irrational case, there are no periodic orbits, so any minimal set M is perfect and its complement is an at most countable collection of open arcs, so we can find $p \in M$ that is not an endpoint of any of these open arcs. Given $\epsilon > 0$, there are positive integers i, j such that the arc $(f^i(p), f^j(p))$ has length at most ϵ and contains p. Let F be a lift of f with p lifting to 0. There are positive integers $k(i)$, $k(j)$ with

$$k(i) - \epsilon < F^i(0) < k(i) \text{ and } k(j) < F^j(0) < k(j) + \epsilon.$$

Thus there are constants $a < 0 < b$ of absolute value less than ϵ such that if $G_\alpha(x) = F(x) + \alpha$, $\alpha = a$ or b, then $G_b^i(0) = k(i)$ and $G_a^j(0) = k(j)$. Clearly each G_α is the lift of a homemorphism g_α of the circle, and the construction makes it clear that

$$\tau(G_a) < \tau(F) < \tau(G_b).$$

For $\alpha = a$, b, let f_α be a diffeomorphism approximating g_α with the same rational rotation number, close enough to g_α that there is a lift F_α of f_α whose graph is ϵ-close to that of F. For $0 \le t \le 1$, $F_t(x) \equiv tF_a(x) + (1-t)F_b(x)$, is smooth, strictly increasing, and satisfies $F_t(x+1) = F_t(x) + 1$, so it is a lift of a diffeomorphism f_t of S^1. By continuity of the rotation number, there is a $t \in (0,1)$ with $\tau(f_t) = \tau(f)$. Since

$$F(x) - 2\epsilon < F_a(x) \le F(x) \le F_b(x) < F(x) + 2\epsilon$$

for all $x \in R$, $d(f, f_t) < 2\epsilon$ as well. □

Recall the definition (5.5) of chain transitivity of a continuous map f on a space X, and in particular that Trans(f) denotes the set of $x \in X$ whose omega-limit sets are all of X.

For $x \in X$, $\epsilon > 0$ and a positive integer n, define

$$G^7(x, n, \epsilon) = \{f \in H(X) : X = \bigcup_{j=0}^{n} V_\epsilon(f^j(x))\}. \tag{9.22}$$

By compactness of X, this is an open subset of $H(X)$, and it follows (see GTDS Theorem 4.12) that if we define:

$$\text{Trans}(X) \underset{\text{def}}{=} \{f \in H(X) : f \text{ is topologically transitive}\}, \text{ then}$$

$$\text{Trans}(X) = \bigcap_{\epsilon} \bigcup_{x,n} G^7(x, n, \epsilon). \tag{9.23}$$

Consequently, Trans(X) is a G_δ subset of $H(X)$. Clearly, Trans(X) is a conjugacy invariant subset of $H(X)$.

By applying some powerful, well known results we can now analyze the irrational case.

THEOREM 9.4. *Let Trans*(S^1) *denote the set of topologically transitive homeomorphisms of* S^1. *Let* c *be an irrational real number.*

(a) *Trans*(S^1) *is a* G_δ *subset of* $H^+(S^1)$ *on which the rotation number function takes irrational values.*

(b) *Let* $f \in H^+(S^1)$ *with* $\tau(f) = c \mod \mathbf{Z}$ *and let* f_c *be the rotation map given by* $f_c(x) = x + c$. $f \in$ *Trans*(S^1) *iff there exists* $h \in H^+(S^1)$ *such that*

$$f = h f_c h^{-1}. \tag{9.24}$$

Thus, Trans$(S^1) \cap \tau^{-1}(c)$ *is the* $H^+(S^1)$ *conjugacy class of* f_c.

(c) *Trans*$(S^1) \cap \tau^{-1}(c)$ *is a* G_δ *set dense in the closed subset* $\tau^{-1}(c) \subset H^+(S^1)$.

PROOF. (a) If $f \in H(S^1)$ and $\text{Per}(f) \neq \emptyset$ then any recurrent point cannot lie in one of the open intervals complementary to $\text{Per}(f)$. That is, $x \in \omega(x, f)$ implies x is periodic. Hence, $\text{Trans}(f) = \emptyset$. Contrapositively, $f \in \text{Trans}(S^1)$ implies $\tau(f)$ is irrational.

(b) This is a classic theorem of Poincaré. See Katok and Hasselblatt (1995) Theorem 11.27.

(c) Given $f \in \tau^{-1}(c)$, apply Lemma 9.3 to choose a C^∞ diffeomorphism f_1 in $\tau^{-1}(c)$ arbitrarily close to f. By a theorem of Denjoy, $f_1 \in \text{Trans}(S^1)$. See Katok and Hasselblatt (1995) Theorem 11.1. Hence, Trans $(S^1) \cap \tau^{-1}(c)$ is dense in $\tau^{-1}(c)$. Since τ is continuous on $H^+(S^1)$, $\tau^{-1}(c)$ is closed and so is G_δ. Hence Trans $(S^1) \cap \tau^{-1}(c)$ is G_δ. □

Thus, in $H(S^1) \backslash H^+(S^1)$ and in each rotation number class $\tau^{-1}(c)$ there is a – necessarily unique – residual $H^+(S^1)$ conjugacy class. For c irrational, it is the class of the rotation f_c which is not only topologically transitive but minimal and equicontinuous. For $H(S^1) \backslash H^+(S^1)$ and for c rational, the classes are those of the complicated maps in $\mathcal{H}_1$.

CHAPTER 10

Crushing the Chain Recurrent Set

Overview: This final crushing argument is an exercise in p.l. topology, to show that for a p.l. manifold a homeomorphism f can be perturbed to get a homeomorphism g whose chain recurrent set can be covered by a collection of disjoint open sets of diameter less than ϵ. The proof of a preliminary topological general position result has a snag in dimension three. The difficulty can be repaired in part by using Moise's Hauptvermutung Theorem for dimension three. Thus, our crushing argument requires that if dimension X equals three or four then $\partial X = \emptyset$.

Our final crushing argument is a version of the initial construction in Shub's proof of his C^0 Density Theorem, Shub (1972). For this foray into piecewise linear topology we review the elements of regular neighborhood theory, following the classic treatment of Cohen (1969).

For a collection of simplices J contained in a simplicial complex K we let $|J|$ denote the union of the open simplices $\overset{\circ}{\sigma}$ with $\sigma < J$. However, we will write σ for a simplex in K, for the subcomplex consisting of this simplex together with its faces and for the associated closed subset of $|K|$. Similarly, we will write $\partial\sigma$ for the complex of proper faces of σ and for the boundary set $\sigma \setminus \overset{\circ}{\sigma}$. For any subset $S \subset |K|$ we define the subcomplexes of K:

$$(10.1)\qquad \begin{cases} N(S,K) = \{\sigma < K : \sigma \text{ is a face of a simplex of } K \text{ meeting } S\} \\ C(S,K) = \{\sigma < K : \sigma \cap S = \emptyset\} \\ \overset{\bullet}{N}(S,K) = N(S,K) \cap C(S,K) \\ \overset{\circ}{N}(S,K) = N(S,K) \setminus \overset{\bullet}{N}(S,K) = \{\sigma < K : \sigma \text{ meets } S\}. \end{cases}$$

A subcomplex K_1 of K is called *full* if $\sigma \cap K_1$ is a face of σ (including the possibilities σ and $\emptyset$) for every simplex σ of K. Any simplex $\sigma < K$ is a full subcomplex. $\partial\sigma$ and the i-skeleton K^i for $i < \dim K$ are not. For any subcomplex K_1 in K the associated *parametrization* is the simplicial map α to a one simplex $[v_1, v]$ sending the vertices of K_1 to v_1 and the remaining vertices of K to v. K_1 is a full subcomplex exactly when $K_1 = \alpha^{-1}(v_1)$. In any case we have

$$(10.2)\qquad \overset{\circ}{N}(|K_1|, K) = \alpha^{-1}([v_1, v] \setminus v).$$

$\alpha^{-1}(v)$ is clearly $C(|K_1|, K)$.

The *first derived subdivision* K' of K is obtained by choosing a new vertex $b(\sigma) \in \overset{\circ}{\sigma}$ for each σ. For each strictly increasing sequence $\sigma_0 < \ldots < \sigma_k$ the convex hull $[b(\sigma_0), \ldots b(\sigma_k)]$ in σ_k is a k simplex of K'. This is called the *barycentric subdivision* of K when each $b(\sigma)$ in the barycenter of σ but as we will see it is often important to make other choices. If we make a different set of choices, $\widetilde{b}(\sigma) \in \overset{\circ}{\sigma}$ to get a different derived subdivision $\widetilde{K}$ then the association $\widetilde{b}(\sigma) \to b(\sigma)$ extends to a simplicial isomorphism h, the *isomorphism relating the deriveds.* Clearly, for each $\sigma < K$ the map h satisfies $h(|\sigma|) = |\sigma|$. For any subcomplex K_1 in K, the derived subdivision K_1' is a full subcomplex of K'.

For $\sigma < K$, the *dual cell* $D(\sigma, K')$ is the subcomplex of the derived consisting of those simplices $[b(\sigma_0) \dots b(\sigma_k)]$ such that $\sigma_0 < \dots < \sigma_k$ with $\sigma < \sigma_0$ or $\sigma = \sigma_0$. Clearly, $|D(\sigma, K')|$ intersects σ in the single point $b(\sigma)$. If K is a p.l. n manifold and σ is an i simplex then $D(\sigma, K')$ is an $n-i$ dimensional p.l. ball. Notice that for the i skeleton K^i:

$$\overset{\bullet}{N}(|K^i|, K') = C(|K^i|, K') = \cup\{D(\sigma, K') : \sigma < K \text{ and } \dim \sigma > i\}. \tag{10.3}$$

This subcomplex of K' is called the *dual* $n-i-1$ *skeleton* where $n = \dim K$.

If K_1 and K_0 are full subcomplexes of K then the complex $N(|K_1| \backslash |K_0|, K')$ is called a *relative regular neighborhood* of $|K_1|$ mod $|K_0|$ in $|K|$ (a *regular neighborhood* of $|K_1|$ in $|K|$ when $K_0 = \emptyset$). If K_1, K_2 and K_0 are full subcomplexes of K with $K_1 \cap K_2 \subset K_0$ then it is easy to check that

$$N(|K_1| \backslash |K_0|, K') \cap N(|K_2| \backslash |K_0|, K') \subset K_1' \cap K_2' \subset K_0'. \tag{10.4}$$

If K is some subdivision of a complex L with $|K| = |L|$ a p.l. n manifold and (K_1, K_0) is the corresponding subdivision of the pair $(\sigma, \partial\sigma)$ for some $\sigma < L$ (or more generally if K_1 is a locally unknotted p.l. ball with boundary K_0) then $N(|K_1| \backslash |K_0|, K')$ is a p.l. n ball with boundary sphere $\overset{\bullet}{N}(|K_1| \backslash |K_0|, K')$ and $(N(|K_1| \backslash |K_0|, K'), |K_1|)$ is an unknotted ball pair.

If K_1 is a full subcomplex of K then a simplex $\sigma < K$ which maps onto $[v_1, v]$ by the parametrization map α can be written as a join $\tau_1\tau$ with $\tau_1 \in K_1$ and $\tau \in C(|K_1|, K)$, neither empty. We call a derived subdivision δ*-thin* for K_1 (or δ*-thick* for K_1) if for each such σ the point $b(\sigma)$ is chosen δ-close to τ_1 (resp. δ-close to τ). By using the parametrization α it is easy to see that for every $\epsilon_1 > 0$ there exists $\delta > 0$ so that if K' and $\widetilde{K}$ are respectively a δ-thin and δ-thick derived subdivision then

$$\begin{cases} N(|K_1|, K') \subset V_{\epsilon_1}(|K|) \\ N(|K_1|, \widetilde{K}) \cup V_{\epsilon_1}(|C(|K_1|, K)|) = |K|. \end{cases} \tag{10.5}$$

For polyhedra X and Y a path of p.l. homeomorphisms $h_t : X \to Y$ $(t \in [0,1])$ is called a *p.l. isotopy* from X to Y (or if $X = Y$ a p.l. isotopy on X) if $(x,t) \mapsto (h_t(x), t)$ defines a p.l. homeomorphism from $X \times I \to Y \times I$. If $h_t = 1_X$ on a subset X_0 of X then h_t is an *isotopy rel* X_0. For $\epsilon > 0$, h_t is an ϵ*-isotropy* if $d(h_0(x), h_t(x)) < \epsilon$ for all $(x,t) \in X \times I$.

The isotopy result we will need is the *Alexander Trick*, see Rourke and Sanderson (1972) p. 37. This says that if B is a p.l. ball and h is a p.l. homeomorphism on B with $h = 1_B$ on ∂B, the boundary sphere, then there exists h_t $(t \in [0,1])$ a p.l. isotopy rel ∂B on B such that $h_0 = 1_B$ and $h_1 = h$. Of course, if the diameter of B is less than ϵ then h_t is an ϵ-isotopy.

The first step in Shub's argument is a little general position move which is easy in the p.l. case. Our homeomorphisms are not p.l. and so some attention must be paid.

LEMMA 10.1. *Let X and Y be p.l. manifolds of dimension $n \neq 3$. Let K, K_0 and L, L_0 be subpolyhedra of X and Y, respectively, with dimension $K \backslash K_0 = k$ and dimension $L \backslash L_0 = l$. Assume that*

$$X \supset K \supset K_0 \supset K \cap \partial X$$

$$Y \supset L \supset L_0 \supset L \cap \partial Y, \tag{10.6}$$

and that the dimensions satisfy

$$k \leq l \text{ and } k + l < n. \tag{10.7}$$

Let $f : X \to Y$ be a homeomorphism. Assume that

$$f(K_0) \cap L = \emptyset \text{ and } f(K) \cap L_0 = \emptyset. \tag{10.8}$$

For any $\epsilon > 0$ there exists a p.l. isotopy $h_t : X \to X$ $(t \in [0,1])$ with $h_0 = 1_X$, such that:

(1) *For all $t \in [0,1]$, $h_t = 1_X$ on $K_0 \cup f^{-1}(L_0) \cup \partial X$ and $d(h_t, 1_X) < \epsilon$.*

(2) *With $g = f \circ h_1$, $g(K) \cap L = \emptyset$.*

PROOF. We use induction on k. When $k = -1$, i.e. $K \backslash K_0 = \emptyset$, let $h = 1_X$.

For the inductive step triangulate X so that we can regard X as a complex with subcomplexes K and K_0. By subdividing we can assume that each simplex of X has diameter less than $\epsilon/2$. Let K^{k-1} be the $k-1$ skeleton of K (empty when $k = 0$). Apply the inductive hypothesis to the pair $(K^{k-1} \cup K_0, K_0)$ to get a p.l. $(\epsilon/2)$-isotopy rel $K_0 \cup f^{-1}(L_0) \cup \partial X$, $h'_t : X \to X$ $(t \in [0,1])$ with $h'_0 = 1_X$ and such that with $f_1 = f \circ h'_1$, we have

$$\begin{cases} f_1(K^{k-1} \cup K_0) \cap L = \emptyset \\ K \cap f_1^{-1}(L_0) = K \cap f^{-1}(L_0) = \emptyset. \end{cases} \tag{10.9}$$

Take a derived subdivision so that σ' and $(\partial\sigma)'$ are full subcomplexes of X' for each $\sigma < X$. For $\sigma < K \backslash K_0$ with dim $\sigma = k$ we denote by B_σ the n ball $|N(|\sigma| \backslash |\partial\sigma|, X'')|$. We can take the second derived δ-thin enough with respect to K so that from the first inclusion in (10.5) we can get

$$\begin{cases} \sigma \cap \partial B_\sigma = \partial\sigma = B_\sigma \cap (K^{k-1} \cup K_0) \\ B_\sigma \cap f_1^{-1}(L_0) = \emptyset \\ \text{diameter } B_\sigma < \epsilon/2. \end{cases} \tag{10.10}$$

From (10.4) it follows that two such balls meet, if at all, in a subset of the boundary spheres. Furthermore, $K \cap \partial X \subset K_0$ and $\sigma < K \backslash K_0$ imply

$$B_\sigma \backslash \partial B_\sigma \subset X \backslash \partial X. \tag{10.11}$$

The pair (B_σ, σ) is a standard p.l. ball pair of dimensions (n, k). Our geometrical work will be directed to the following:

Claim: For each σ, there exists a k dimensional p.l. ball $\widetilde{B}_\sigma \subset B_\sigma$ with $\widetilde{B}_\sigma \cap \partial B_\sigma = \partial \widetilde{B}_\sigma = \partial\sigma$ and such that $\widetilde{B}_\sigma \cap f_1^{-1}(L) = \emptyset$. Furthermore, the ball pair $(B_\sigma, \widetilde{B}_\sigma)$ is unknotted.

To say that the pair $(B_\sigma, \widetilde{B}_\sigma)$ is unknotted is exactly to say that there is a p.l. homeomorphism h_σ from (B_σ, σ) to $(B_\sigma, \widetilde{B}_\sigma)$. By preceding by the inverse of the cone of the map on the boundary sphere pair, we can ensure that $h_\sigma = 1_{B_\sigma}$ on ∂B_σ. Then define h_1'' to be h_σ on B_σ for each such $\sigma < K\backslash(K_0 \cup \partial X)$ and to be the identity elsewhere. By using the Alexander Trick on each ball we get a p.l. isotopy h_t'', $t \in [0,1]$ which is the identity for all t outside the balls B_σ. In particular, the isotopy is rel

$$K^{k-1} \cup K_0 \cup f_1^{-1}(L_0) \cup \partial X = K^{k-1} \cup K_0 \cup f^{-1}(L_0) \cup \partial X.$$

Because the diameter of each B_σ is less than $\epsilon/2$, h_t'' is an $(\epsilon/2)$-isotopy.

Let $h_1 = h_1' \circ h_1''$. On $K^{k-1} \cup K_0 \cup f^{-1}(L_0)$, $g = f \circ h_1 = f \circ h_1' = f_1$ and so by (10.9) the image of this set under g is disjoint from L. For each $\sigma < K\backslash(K_0 \cup K^{k-1})$ $g(\sigma) = f_1 h_1''(\sigma) = f_1(\widetilde{B}_\sigma)$ which is disjoint from L. It follows that the required isotopy is given by

$$h_t = \begin{cases} h_{2t}' & 0 \le t \le \frac{1}{2} \\ h_1' \circ h_{2t-1}'' & \frac{1}{2} \le t \le 1. \end{cases} \tag{10.12}$$

To prove the Claim we will approximate the embedding $f_1|\sigma : \sigma \to Y$ by a p.l. map, use general position to move the new map off L and then pull back via f_1 to get a map of σ into $B_\sigma \backslash f_1^{-1}(L)$ which is the inclusion on $\partial\sigma$. Then make this map p.l. and use general position in X to get a p.l. embedding.

In detail, we think of the simplex σ as made up of two pieces, a slightly smaller dimension k simplex, $\sigma' \subset \overset{\circ}{\sigma}$, and a collar which we think of as $\partial\sigma \times [0,1]$, identifying $\partial\sigma \times 0$ with $\partial\sigma$ and $\partial\sigma \times 1$ with $\partial\sigma'$. By (10.9) $f_1(\partial\sigma) \cap L = \emptyset$ and so we can choose the collar thin enough that $f_1(\partial\sigma \times [0,1]) \cap L = \emptyset$. Because f_1 is a homeomorphism $f_1(B_\sigma^\circ)$ is an open subset of Y which contains $f_1(\partial\sigma \times (0,1] \cup \sigma') = f_1(\overset{\circ}{\sigma})$.

The map f_1 on σ is not p.l. but by the p.l. approximation theorem we find a p.l. map on σ' close to f_1 on σ' and extend the new map on $\partial\sigma'$ to a little homotopy on the $[1/2, 1]$ piece of the collar. In sum we can obtain a continuous map $q_1 : \sigma \to Y$ which satisfies:

$$\begin{cases} q_1 = f_1 \text{ on } \partial\sigma \times [0,1/2] \\ q_1(\partial\sigma \times [0,1]) \cap L = \emptyset \\ q_1(\partial\sigma \times [1/2,1] \cup \sigma') \subset f_1(B_\sigma^\circ) \\ q_1 : \sigma' \to Y \text{ is a p.l map.} \end{cases} \tag{10.13}$$

Notice that the map q_1 need no longer be injective on σ. Nonetheless q_1 is a p.l. map on σ' with $q_1(\partial\sigma') \cap L = \emptyset$ and $q_1(\sigma') \cap L_0 \subset f_1(B_\sigma^\circ) \cap L_0 = \emptyset$. So $q_1(\sigma')$ and $\overline{L\backslash L_0}$ are subpolyhedra of Y of dimensions at most k and l, with $k + l \le n - 1$. So by general position for subsets (see Rourke and Sanderson (1972) Theorem 5.3) we can isotope $q_1(\sigma')$ off $\overline{L\backslash L_0}$ rel $q_1(\partial\sigma')$. Composing q_1 with the p.l. homeomorphism doing the moving we get q_2 on σ', agreeing with q_1 on $\partial\sigma'$ and such that $q_1(\sigma') \cap \overline{L\backslash L_0} = \emptyset$. We can make the move small enough that $q_2(\sigma')$ remains in $f_1(B_\sigma^\circ)$. Extend q_2 to agree with q_1 on the collar, and define $q_3 = f_1^{-1} \circ q_2$. Thus, q_3 is a continuous map from $\sigma = \partial\sigma \times [0,1] \cup \sigma'$ to X which satisfies:

$$(10.14)\qquad \begin{cases} q_3 = 1_\sigma \text{ on } \partial\sigma \times [0,1/2] \\ \\ q_3(\sigma) \cap f_1^{-1}(L) = f_1^{-1}(q_2(\sigma) \cap L) = \emptyset \\ \\ q_3(\partial\sigma \times [1/2,1] \cup \sigma') \subset B_\sigma^\circ. \end{cases}$$

The map $q_3 : \sigma \to X$ is not p.l. but its restriction to $\partial\sigma \times [0,1/2]$ is p.l. So by the relative p.l. approximation theorem there is a p.l. map $q_4 : \sigma \to X$ which agrees with q_3 on $\partial\sigma \times [0,1/2]$. We can choose the approximation close enough that q_4 maps $\partial\sigma \times [1/2,1] \cup \sigma'$ into the open set $(B_\sigma \backslash f_1^{-1}(L))^\circ$ just as q_3 did. So we have obtained a p.l. map $q_4 : \sigma \to X$ with $q_4 = 1_\sigma$ on $\partial\sigma \times [0,1/2]$ and $q_4(\sigma) \subset (B_\sigma \backslash f_1^{-1}(L))^\circ$. Now $2\dim\sigma + 1 = 2k+1 \le k+l+1 \le n$ and so by general position for p.l. maps we can replace q_4 by a p.l. embedding q_5 which agrees with q_4 on $\partial\sigma \times [0,1/2]$ and which still maps $\partial\sigma \times [1/2,1] \cup \sigma'$ to $(B_\sigma \backslash f_1^{-1}(L))^\circ$. The disk required by the Claim is $\widetilde{B}_\sigma = q_5(\sigma)$.

The unknotting result is obvious when $k = 0$ in which case σ and $\widetilde{B}_\sigma$ are points in the interior of B_σ. Otherwise, $k \le l$ and $k + l \le n-1$ imply that $k \le [(n-1)/2]$. If $n \ge 4$ then $[(n-1)/2] \le n-3$ and if $n \le 2$ then $[(n-1)/2] \le 0$. So if $k > 0$ and $n \ne 3$ the ball pair has codimension at least 3 and so is unknotted, see Rourke and Sanderson (1972) Theorem 7.1. This argument fails when $k = l = 1$ and $n = 3$. □

PROPOSITION 10.2. *Let X and Y be p.l. manifolds of dimension n. Assume that either $\partial X = \partial Y = \emptyset$ or $n \ne 3, 4$. Let $K_0 \subset K_1 \subset \ldots \subset K_r \subset X$ and $Y \supset L_0 \supset L_1 \ldots \supset L_r$ be subpolyhedra such that*

$$(10.15)\qquad \begin{cases} \dim K_i + \dim L_i < n \\ \\ \dim(K_i \cap \partial X) + \dim(L_i \cap \partial Y) < n-1 \end{cases}$$

for $i = 0, \ldots, r$. Let $f : X \to Y$ be a homeomorphism. Let $\epsilon > 0$.

There exists a homeomorphism $g : X \to Y$ with $d(f,g) \le \epsilon$ satisfying $g(K_i) \cap L_i = \emptyset$ for $i = 0, 1, \ldots, r$.

PROOF. First assume that $\partial X = \partial Y = \emptyset$. We separate the $n = 3$ case, assuming first that $n \ne 3$. Let r_0 be the largest index such that $\dim K_i \le \dim L_i$ for all $i \le r_0$. Let δ_1 be an $\epsilon/2$ modulus of uniform continuity for f. Apply the lemma successively to move K_0 off L_0, K_1 off L_1 up through K_{r_0} and L_{r_0}. Thus, we obtain $h_1 : X \to X$ with $d(h_1, 1_X) \le \delta_1$ such that with $f_1 = f \circ h_1$, $f_1(K_i) \cap L_i = \emptyset$ for $0 \le i \le r_0$. By definition of δ_1, $d(f, f_1) \le \epsilon/2$. Now let δ_2 be less than $\epsilon/2$ and such that $\overline{V_{\delta_2}(f_1(K_i))} \cap L_i = \emptyset$ for $0 \le i \le r_0$. Apply the lemma to f_1^{-1} to move L_r off K_r, down through L_{r_0+1} and K_{r_0+1}, obtaining $h_2 : Y \to Y$ with $d(h_2, 1_X) \le \delta_2$ so that $(f_1^{-1} \circ h_2)(L_i) \cap K_i = \emptyset$ for $r_0 < i \le r$. So with $g = h_2^{-1} \circ f_1 = h_2^{-1} \circ f \circ h_1$ we have $L_i \cap g(K_i) = \emptyset$ for $r_0 < i \le r$ and $g(K_i) \cap L_i \subset V_{\delta_2}(f_1(K_i)) \cap L_i = \emptyset$ for $0 \le i \le r_0$.

There remains the case n = 3. With some embarrassment we deploy the nuclear weaponry of Moise's Hauptvermutung results. Because $\partial X = \partial Y = \emptyset$, we can apply Theorem 2 of Moise (1952) to $(\epsilon/2)$-approximate the homeomorphism $f : X \to Y$ by a p.l. homeomorphism $f_1 : X \to Y$. Then successive applications of ordinary general position for subpolyhedra allows us to move $f_1(K_i)$ off L_i for $0 \le i \le r$.

That is, we obtain $h : Y \to Y$ with $d(h, 1_Y) < \epsilon/2$ so that $hf_1(K_i) \cap L_i = \emptyset$ for all i.

Now assume that $\partial X, \partial Y \neq \emptyset$ but $n \neq 3, 4$. First apply the previous result to the restriction of f to the boundaries. Because $n - 1 \neq 3$ the lemma yields not only homeomorphisms but isotopies. Thus, we can construct $f_0 : \partial X \to \partial Y$ with $f_0(K_i \cap \partial X) \cap L_i \cap \partial Y = \emptyset$ for $i = 0, \dots, r$ and such that f_0 is $(\epsilon/2)$-isotopic to $f : \partial X \to \partial Y$. Because the boundary of a p.l. manifold is collared we can extend f_0 to a homeomorphism $f_0 : X \to Y$ which equals f on the complement of some neighborhood of ∂X and with $d(f, f_0) < \epsilon/2$. Because $n \neq 3$ we can apply Lemma 10.1 as in the previous argument to move K_i off L_i using homeomorphisms h which restrict to the identity of ∂X and ∂Y. □

Jerome Dancis pointed out to us that the codimension three results can be obtained by using the Taming Lemma to play the role Moise's result did in dimension three. The Taming Lemma is described in Dancis (1976) as the concatenation of results of Bryant, Seebeck, Cernavskii, Homma and Miller. We included the direct proof of Lemma 10.1 to keep the exposition as self-contained as possible.

PROPOSITION 10.3. *Let f be a homeomorphism on X with X homeomorphic to a piecewise linear manifold of dimension n. Assume that either $\partial X = \emptyset$ or $n \neq 3, 4$. Let $\epsilon > 0$.*

There exists $\{O_1, O_2, \dots, O_r\}$ a pairwise disjoint family of open subsets of X and $g \in H(X)$ such that:

(1) *$d(f, g) < \epsilon$ and the diameter of O_i is less than ϵ for $i = 1, \dots, r$.*

(2) *The chain recurrent set $\mathcal{C}(g)$ is contained in the union $\cup_{i=1}^r O_i$.*

PROOF. We can triangulate and subdivide to assume that $X = |K|$ is a p.l. manifold of dimension n with each simplex $\sigma < K$ of diameter less than $\epsilon/2$. Let K' be the barycentric derived subdivision of K.

We first move the image of the i skeleton off the dual of the i skeleton for $i = 0, \dots, n-1$. That is, by (10.3), $\dim K^i + \dim C(|K^i|, K') = n - 1$ and so by Corollary 10.2 there exists $f_1 \in H(X)$ such that $d(f, f_1) < \epsilon/2$ and such that $f_1(|K^i|) \cap C(|K^i|, K') = \emptyset$ for $i = 0, \dots, n-1$. As these are disjoint closed sets there exists $\epsilon_1 > 0$ such that

$$f_1(\overline{V_{\epsilon_1}(|K^i|)}) \cap \overline{V_{\epsilon_1}(|C(|K^i|, K')|)} = \emptyset \tag{10.16}$$

for $i = 0, 1, \dots n - 1$.

Now we go to the second derived subdivision making two different choices. Given $\delta > 0$, we choose for the simplex of K' $[b(\sigma_0), \dots, b(\sigma_k)]$ (with $\sigma_0 < \dots < \sigma_k$ in K), the *thin* choice of interior point is δ-close to $b(\sigma_0)$ while the *thick* choice is δ-close to $b(\sigma_k)$. Let L denote the second derived K'' with the thin choice and $\widetilde{L}$ the second derived K'' with the thick choice and let $h : \widetilde{L} \to L$ be the simplicial isomorphism associating the two deriveds. In particular, each simplex $[b(\sigma_0), \dots, b(\sigma_k)]$ of K' is invariant under h and so, since σ_k has diameter less than $\epsilon/2$ we have

$$d(h, 1_X) \leq \epsilon/2. \tag{10.17}$$

Our thin (or thick) choices are thin (resp. thick) for each full subcomplex $(K^i)'$ of K' $(i = 0, 1, \dots, i)$, and for each σ' of K' where $\sigma < K$. So by (10.5) we can choose $\delta > 0$ small enough that for $i = 0, 1, \ldots, n-1$:

$$N(|K^i|, L) \subset V_{\epsilon_1}(|K^i|)$$

$$N(|K^i|, \widetilde{L}) \supset |K| \backslash \overline{V_{\epsilon_1}(|C(|K^i|, K')|)}, \tag{10.18}$$

and furthermore that for each $\sigma < K$:

$$\text{diameter } N(|\sigma|, L) < \epsilon. \tag{10.19}$$

Let $U_i = |N(|K^i|, L)|$ $(i = 0, \ldots, n)$ and let $g = h \circ f_1$. The isomorphism h is our crushing map. By (10.17) and (10.19) f_1 maps each U_i into the interior of $|N(|K^i|, \widetilde{L})|$. Then h maps this thick neighborhood isomorphically back onto the thin neighborhood U_i. Hence, each U_i is an inward set for g. Clearly, $\emptyset \equiv U_{-1} \subset U_0 \subset U_1 \subset \ldots \subset U_{n-1} \subset U_n \equiv X$.

Because the U_i's are inward, Lemma 1.5 shows that any chain component B of g which meets $U_i \backslash (U_{i-1}^{\circ})$ is contained in the open set $U_i^{\circ} \backslash U_{i-1}$, which we will call $\widetilde{O}_i$. It follows that the chain recurrent set $\mathcal{C}(g)$ is contained in the union of the sequence of disjoint open sets $\{\widetilde{O}_0, \widetilde{O}_1, \ldots, \widetilde{O}_n\}$. We complete the proof by showing that each $\widetilde{O}_i$ is the disjoint union of open sets of diameter less than ϵ.

The set $U_i = |N(|K^i|, L)|$ is the union of the regular neighborhood $U_{i-1} = |N(|K^{i-1}|, L)|$ and the relative regular neighborhoods $\{|N(|\sigma| \backslash |\partial\sigma|, L)| : \sigma < K$ with $\dim \sigma = i\}$. Furthermore, if σ_1, σ_2 are distinct simplices of dimension i then by (10.4)

$$|N(|\sigma_1| \backslash |\partial\sigma_1|, L)| \cap |N(|\sigma_2| \backslash |\partial\sigma_2|, L)| \subset |\sigma_1| \cap |\sigma_2|, \tag{10.20}$$

and so the intersection is contained in $|K^{i-1}| \subset U_{i-1}$. Thus, $\widetilde{O}_i$ is the disjoint union of the open sets $(|N(|\sigma| \backslash |\partial\sigma|, L|)^{\circ} \backslash U_{i-1}$ as σ varies over the K simplices of dimension i. By (10.20) each of these sets has diameter less than ϵ. □

CHAPTER 11

Generic Homeomorphisms on Manifolds

Overview: Here we describe how our earlier results yield the consequences for homeomorphisms on manifolds that we outlined in the Introduction.

Throughout this section X is a space homeomorphic to a p.l. manifold of dimension n and $H = H(X)$ is the automorphism group acting on X and ∂X by evaluation. $\mathcal{U}_B$ is the base of balls defined by (2.21). X is generalized homogeneous on $X \backslash \partial X$ if $n \geq 2$. By Proposition 4.6 $n \geq 2$ implies X is spongy and if $n > 2$ or $n = 2$ with $\partial X = \emptyset$ then X is strictly spongy.

THEOREM 11.1. *Let X be homeomorphic to a piecewise linear manifold of dimension n. Assume that $n \geq 2$ and either $n > 2$ or $\partial X = \emptyset$.*

The sets $\mathcal{H}_{1,s}[\mathcal{U}_B]$, $\mathcal{H}_{3,s}[\mathcal{U}_B]$, $\mathcal{H}_4$ defined by (6.6), (6.15) *and* (6.29) *respectively are dense, conjugacy invariant, G_δ subsets of $H(X)$ as are the sets for which the restricted action of $H(X)$ on ∂X satisfies the corresponding properties.*

If $f \in \mathcal{H}_{3,s}[\mathcal{U}_B] \cap \mathcal{H}_4$ then f satisfies the properties listed in Theorems 6.3, 6.4, 6.8, 6.9, 6.13 and 7.6 together with Corollaries 6.5 and 6.10. (*For the most part, these are the properties* (1) *through* (5) *that were enumerated in the Introduction.*)

PROOF. That the sets are conjugacy invariant dense, G_δ subsets follows from Theorems 6.2, 6.7 and 6.12. In Theorem 6.7 it was also shown that $\mathcal{H}_{3,s}[\mathcal{U}_B] \subset \mathcal{H}_{1,s}[\mathcal{U}_B]$. In particular, the listed results apply to $f \in \mathcal{H}_{3,s}[\mathcal{U}_B] \cap \mathcal{H}_4$. □

11.1. The class $\mathcal{H}_8$

We call a subset $S \subset X$ *ϵ-thin* if it can be covered by a family of pairwise disjoint open sets, each of diameter $< \epsilon$. For $\epsilon > 0$ define the subset $G^8(\epsilon)$ of $H(X)$ by

$$f \in G^8(\epsilon) \Leftrightarrow \mathcal{C}(f) \text{ is } \epsilon\text{-thin.} \tag{11.1}$$

LEMMA 11.2. *For any compact metric space Y, each $G^8(\epsilon)$ is an open subset of $H(Y)$. If X is homeomorphic to a p.l. manifold of dimension n and either $\partial X = \emptyset$ or $n \neq 3, 4$ then $G^8(\epsilon)$ is dense in $H(X)$ for every $\epsilon > 0$.*

PROOF. The association $f \mapsto \mathcal{C}(f)$ is an upper semicontinuous relation from $H(X)$ to X for any space X, see GTDS Theorem 7.23. It follows that if $\mathcal{C}(f) \subset O$ for some open set O which is a disjoint union of open sets of small diameter then $\mathcal{C}(f_1) \subset O$ for all f_1 close enough to f. It obviously follows that $G^8(\epsilon)$ is open. Density follows from the hypotheses by Proposition 10.3. □

Define

$$\mathcal{H}_8 \underset{\text{def}}{=} \bigcap_{\epsilon > 0} G^8(\epsilon). \tag{11.2}$$

THEOREM 11.3. *For any space X, $\mathcal{H}_8$ is a conjugacy invariant, G_δ subset of $H(X)$ and $f \in \mathcal{H}_8$ iff $\mathcal{C}(f)$ is zero-dimensional.*

If X is homeomorphic to a p.l. manifold of dimension n and either $\partial X = 0$ or $n \neq 3, 4$ then $\mathcal{H}_8$ is dense in $H(X)$.

PROOF. To get $\mathcal{H}_8$ we can intersect over positive rational ϵ and so $\mathcal{H}_8$ is a G_δ by Lemma 11.2. Any closed subset A of X is zero-dimensional iff for every $\epsilon > 0$ it can be covered by a family of pairwise disjoint open sets of diameter $< \epsilon$. Hence, $\mathcal{C}(f)$ is zero-dimensional iff $f \in \mathcal{H}_8$. This condition is clearly conjugacy invariant. Finally, density follows in the manifold cases by Lemma 11.2 again. □

PROPOSITION 11.4. *Let X be a locally connected space, e.g. a manifold, and let f be a homeomorphism on X. If $f \in \mathcal{H}_8$ then a chain component B for f is terminal iff $B \subset \Pi_\infty(f)$ and so is a periodic orbit or an adding machine.*

PROOF. $B \subset \mathcal{C}(f)$ is zero-dimensional and X is locally connected. The result then follows from Proposition 2.3. □

We use this result to sharpen Theorem 7.5:

PROPOSITION 11.5. *Let X be a locally connected space, e.g. a manifold, and let f be a homeomorphism on X. If f is in the residual set $\mathcal{H}_{3,s} \cap \mathcal{H}_8$ then the set of chain continuity points of f is*

$$W^s_f(\Pi_\infty(f)) = \{x \in X : \omega(x, f) = \omega\mathcal{C}(x, f)\}. \tag{11.3}$$

PROOF. Since $f \in \mathcal{H}_{3,s}$, we can apply Proposition 3.6 to see that $\omega(x, f) = \omega\mathcal{C}(x, f)$ iff $\omega(x, f)$ is a terminal chain component. The result now follows from Theorem 7.5 and Proposition 11.4. □

11.2. The class $\mathcal{H}_{man}$

Define

$$\mathcal{H}_{man}(X) = \mathcal{H}_{3,s}[\mathcal{U}_B] \cap \mathcal{H}_4 \cap \mathcal{H}_8. \tag{11.4}$$

From Theorems 11.1 and 11.3 we have

THEOREM 11.6. *Let X be a space homeomorphic to a piecewise linear manifold of dimension n. Assume that $n \geq 2$ and that either $\partial X = \emptyset$ or $n \neq 3, 4$. $\mathcal{H}_{man}(X)$ is a dense, conjugacy invariant, G_δ subset of $H(X)$. If $\partial X \neq \emptyset$ and $n \geq 5$ then $\{f \in \mathcal{H}_{man}(X) : f|\partial X \in \mathcal{H}_{man}(\partial X)\}$ is a dense, conjugacy invariant, G_δ subset of $H(X)$.*

PROOF. For the boundary results, we observe that $n \geq 3$ implies that the action of $H(X)$ on ∂X is strictly spongy and generalized homogeneous. Hence, we can apply Theorems 6.2, 6.7 and 6.12 to this action as well. For $\mathcal{H}_8$ notice that for any closed, f-invariant subset A of X, $\mathcal{C}(f|A) \subset \mathcal{C}(f)$. Hence, if $\mathcal{C}(f)$ is zero-dimensional, then so is $\mathcal{C}(f|\partial X)$. □

11.3. Anosov homeomorphisms

For the remainder of the section, assume that X is a smooth manifold of dimension $n \geq 2$ with $\partial X = \emptyset$. Since such a manifold is smoothly triangulable X is homeomorphic to a p.l. manifold. We turn our attention to what our generic homeomorphisms look like in the neighborhood of some Anosov diffeomorphism. We begin with some definitions and background.

A homeomorphism g on a space Y is called *expansive* if for some positive ϵ^*

$$x_1, x_2 \in Y \text{ and } d(g^i(x_1), g^i(x_2)) \leq \epsilon^* \quad (i \in \mathbf{Z}) \Leftrightarrow x_1 = x_2. \tag{11.5}$$

In that case, $\epsilon^* > 0$ is called an *expansivity constant.*

The homeomorphism g is said to satisfy the *Shadowing Property* if for every $\epsilon > 0$ there exists $\delta > 0$ such that every δ-chain for g can be ϵ-shadowed by a g-orbit. That is, $\{x_0, \dots, x_n\}$ with $d(g(x_{i-1}), x_i) < \delta$ for $i = 1, \dots, n$ implies there exists $x \in Y$ such that $d(g^i(x), x_i) < \epsilon$ for $i = 1, \dots, n$.

A homeomorphism g is called an *Anosov homeomorphism* if it is expansive and satisfies the Shadowing Property. For an exposition see GTDS Chapter 11 and Aoki and Hirade (1994) passim; some of the basic facts can be found in Walters (1978).

Thus, $g \in H(Y)$ is an Anosov homeomorphism if for every $\epsilon^* > 0$ sufficiently small, there exists $\delta^* > 0$ such that if $\{x_i : i \in \mathbf{Z}\}$ is a bi-infinite sequence in Y with

$$d(g(x_{i-1}), x_i) \leq \delta^* \quad \text{for all } i \in \mathbf{Z}, \tag{11.6}$$

then there exists a unique $x \in Y$ such that

$$d(g^i(x), x_i) < \epsilon^* \quad \text{for all } i \in \mathbf{Z}. \tag{11.7}$$

The shift homeomorphism σ of (5.4) on Σ is an Anosov homeomorphism. An Anosov diffeomorphism on the smooth manifold X, which satisfies a hyperbolicity condition on all of X, is an Anosov homeomorphism, see e.g. GTDS Theorem 11.29. The simplest example is the Thom torus map, see GTDS pp. 245-246. We use a number of well-known facts about Anosov diffeomorphisms below; general references for these are Robinson (1995) and Katok and Hasselblatt (1995).

An Anosov homeomorphism has only countably many periodic points. In fact, this follows just from expansiveness.

LEMMA 11.7. *If g is an expansive homeomorphism then for every positive integer k the set of periodic points of period k is finite and so* Per(g) *is countable.*

PROOF. Let $\epsilon^* > 0$ be an expansivity constant for g and let $\delta_k > 0$ be an ϵ^* modulus of uniform continuity for $\{g^j : j = 0, \dots, k-1\}$. If x_1, x_2 are fixed by g^k with $d(x_1, x_2) < \delta_k$ then $d(g^j(x_1), g^j(x_2)) < \epsilon^*$ for $j = 0, \dots, k-1$ and so by periodicity for all $j \in \mathbf{Z}$. Hence, $x_1 = x_2$. As the set Fix(g^k) is discrete, it is finite by compactness. □

As the notation suggests, an Anosov diffeomorphism g on X is an example of an Anosov homeomorphism (see Robinson (1995) or Katok and Hasselblatt (1996) for a detailed definition of an Anosov diffeomorphism; it is more than an expansive diffeomorphism with the shadowing property). As is well known (see below for details) if f is a homeomorphism on X which is close enough to g in the C^0, uniform, metric then there is a surjective continuous map h on X mapping f to g, i.e. there is a semi-conjugacy from f to g. This last property is sometimes called *topological stability* of g, although this terminology has other meanings in other contexts. A proof that an Anosov homeomorphism on a compact manifold is topologically stable can be found in Walters (1978). Theorem 11.8 below describes the structure of this map h when f is one of the generic homeomorphisms described above. It is quite degenerate, as one would expect after contemplating the great differences in the dynamics of f and of g – g is expansive and has the entire manifold as a single chain component, while f is almost equicontinuous and has uncountable many distinct chain components whose union is a Cantor set.

THEOREM 11.8. *Let X be a smooth, connected manifold of dimension $n \geq 2$ with $\partial X = \emptyset$. Let g be a topologically transitive Anosov homeomorphism on X. Assume that $\delta^* < \epsilon^*$ are positive reals such that $3\epsilon^*$ is an expansivity constant for g and every δ^* chain for g can be ϵ^* shadowed by the g orbit of a unique point in X. Assume that ϵ^* is small enough that any continuous map on X which is ϵ^* close to the identity is surjective. Let f be a homeomorphism on X with $d(f,g) \leq \delta^*$.*

There exists a unique continuous, surjective map h on X which satisfies

$$h \circ f = g \circ h \tag{11.8}$$

and

$$d(h, 1_X) \leq \epsilon^*. \tag{11.9}$$

If $f \in \mathcal{H}_{man}(X)$ then the semiconjugacy map h has the following properties.

(1) h maps each terminal chain component for f and each initial chain component for f to a periodic orbit of g, i.e

$$h(\Pi_\infty(f) \cup \Pi_\infty(f^{-1})) \subset \mathrm{Per}(g). \tag{11.10}$$

(2) There is an open, dense, f-invariant subset O of X on which h is locally constant. The image set $h(O)$ is a countable set with

$$h(O) \subset W^s_g(\mathrm{Per}(g)) \cap W^u_g(\mathrm{Per}(g)). \tag{11.11}$$

That is, if $x \in O$ then the g orbit of $h(x)$ is asymptotic to periodic orbits in both the positive and negative directions.

(3) There exists a chain component B for f, such that h maps B onto all of X, i.e.

$$h(B) = X. \tag{11.12}$$

B is a zero-dimensional subset of X which is ϵ^* dense in X, i.e.

$$\overline{V_{\epsilon^*}(B)} = X. \tag{11.13}$$

PROOF. The construction of h can be found in Walters (1978); here is an outline. Since $d(f,g) \leq \delta^*$ the bi-infinite f orbit sequence of x is a δ^* chain for g. So it is ϵ^* shadowed by the g orbit of a unique point $h(x)$. The f orbit of $f(x)$ is ϵ^* shadowed by the g orbit of $g(h(x))$ and so by uniqueness $h(f(x)) = g(h(x))$. For the proof that h is continuous and surjective, see Walters (1978).

Conversely, if $h : X \to X$ satisfies (11.9) then $d(h(f^i(x)),\ f^i(x)) \leq \epsilon^*$ for all $i \in \mathbf{Z}$. If h also satisfies (11.8) then $h(f^i(x)) = g^i(h(x))$ for all $i \in \mathbf{Z}$. That is, the g-orbit of $h(x)$ ϵ^*-shadows the f-orbit of x. Hence, conditions (11.8) and (11.9) characterize the map h.

Now assume that U is an inward set for f with a (δ^*, k)-periodic decomposition $\{U_i\}$. Let $A = \omega(U, f)$ be the associated attractor with periodic decomposition

$\{U_i \cap A\}$, extended periodically for all $i \in \mathbf{Z}$ as in (2.2). If $x \in U_0 \cap A$ then $f^i(x) \in U_i \cap A$ for all $i \in \mathbf{Z}$. Hence, if $x_1 \in U_0 \cap A$ then

$$\text{(11.14)} \quad \begin{cases} d(g^i(h(x)), g^i(h(x_1))) \\ \leq d(g^i(h(x)), f^i(x)) + d(f^i(x), f^i(x_1)) + d(f^i(x_1), g^i(h(x_1))) \\ \leq 3\epsilon^*. \end{cases}$$

Since $3\epsilon^*$ is an expansivity constant for g we have

$$\text{(11.15)} \quad h(x) = h(x_1).$$

In particular, if we set $x_1 = f^k(x)$ then $x_1 = f^k(x) \in U_k \cap A = U_0 \cap A$

$$\text{(11.16)} \quad h(x) = h(f^k(x)) = g^k(h(x)).$$

Thus, h maps $U_0 \cap A$ to a single periodic point for g with period k, and so g maps A to a single periodic orbit.

If $x \in \Pi_\infty(f)$ then x is contained in some (δ^*, k)-periodic inward set U and, by Proposition 2.2, x lies in the attractor $\omega(U, f)$. Hence, $h(x) \in \mathrm{Per}(g)$. Similarly, if $x \in \Pi_\infty(f^{-1})$ then $h(x) \in \mathrm{Per}(g)$. This proves (11.10) and so property (1) holds because for $f \in \mathcal{H}_8$ every initial and terminal chain component lies in $\Pi_\infty(f) \cup \Pi_\infty(f^{-1})$ by Proposition 11.4.

Now define

$$\text{(11.17)} \quad \begin{cases} O^s = \bigcup\{f^{-j}(U^\circ) : U \text{ a } (\delta^*, k) - \text{periodic inward set for } f \text{ and } j, k \text{ positive integers}\}. \\ \\ O^u = \bigcup\{f^j(\widehat{U}^\circ) : \widehat{U} \text{ a } (\delta^*, k) - \text{periodic inward set for } f^{-1} \text{ and } j, k \text{ positive integers}\}. \end{cases}$$

Let $x \in O \underset{\text{def}}{=} O^s \cap O^u$. There exists (δ^*, k)-periodic inward set U for f, a (δ^*, k')-periodic inward set $\widehat{U}$ for f^{-1}, a positive integer j, and a positive number $\delta_1 < \delta^*$ such that – after possible renumbering of the decompositions for U and $\widehat{U}$:

$$\text{(11.18)} \quad \begin{cases} V_{\delta_1}(f^j(x)) \subset U_0^\circ \\ V_{\delta_1}(f^{-j}(x)) \subset \widehat{U}_0^\circ \end{cases}$$

Now let $\delta_2 > 0$ be a δ_1 modulus of continuity for $\{f^i : |i| \leq j\}$ at x. That is, if $x_1 \in V_{\delta_2}(x)$ then

$$\text{(11.19)} \quad \begin{cases} d(f^i(x), f^i(x_1)) & \leq & \delta_1 \leq \delta^* \text{ for } |i| \leq j \\ \{f^j(x), f^j(x_1)\} & \subset & U_0 \\ \{f^{-j}(x), f^{-j}(x_1)\} & \subset & \widehat{U}_0. \end{cases}$$

It follows that for every positive integer l

$$\text{(11.20)} \quad \begin{cases} \{f^{j+l}(x), f^{j+l}(x_1)\} \subset U_l \\ \\ \{f^{-(j+l)}(x), f^{-(j+l)}(x_1)\} \subset \widehat{U}_l. \end{cases}$$

Since the diameters of the U_l's and U_l''s are at most δ^* we have from (11.20) and (11.19) that

$$\text{(11.21)} \quad d(f^i(x), f^i(x_1)) \leq \delta^* \text{ for all } i \in \mathbf{Z}.$$

It follows that (11.14) and so (11.15) hold. That is, h is constant on $V_{\delta_2}(x)$.

Clearly, O^s and O^u are open. Furthermore,

$$\text{(11.22)} \quad W^s_f(\Pi_\infty(f)) \subset O^s \text{ and } W^u_f(\Pi_\infty(f^{-1})) \subset O^u.$$

For $f \in \mathcal{H}_{man}\mathcal{H}_{3,s} \subset \mathcal{H}_1$, Theorem 6.4(b) shows that O^s and O^u are dense in X. Each is clearly f-invariant. Hence, the intersection O is open, dense and f-invariant. We have just seen that h is locally constant on O. Hence, the image $h(O)$ is the same as the image of the restriction of h to a countable dense subset of O and so $h(O)$ is countable.

With x, U and $\widehat{U}$ as above $\omega(x, f)$ is contained in the attractor $\omega(U, f)$ and so $\omega(h(x), g) = h(\omega(x, f)) \subset h(\omega(U, f))$ which is a periodic orbit. It follows that the g-orbit of $h(x)$ is asymptotic to this periodic orbit. Similarly, the g-orbit of $h(x)$ is asymptotic in the negative direction to the periodic orbit of g^{-1} which is $h(\alpha(\widehat{U}, f))$.

To prove (11.12) we recall the argument at the end of the proof of Proposition 5.2. Consider the class of closed, f-invariant subsets of X which are mapped by h onto X. X is in this class. By the usual Zorn's Lemma argument we can choose a minimal member of this class; call it A. Because g is topologically transitive we can choose $x \in A$ such that $h(x) \in \text{Trans}(g)$. Because A is closed and invariant, $\omega(x, f)$ is a closed f-invariant subset of A and

$$h(\omega(x, f)) = \omega(h(x), g) = X. \tag{11.23}$$

By minimality, $\omega(x, f) = A$. That is, x is a transitive point for the restriction of f to A.

Since f is topologically transitive on A, A is a chain transitive subset of X and so A is contained in some chain component B. Since B contains A, (11.12) holds.

Because $f \in \mathcal{H}_8$ the whole chain recurrent set $\mathcal{C}(f)$ is zero-dimensional and so B is zero dimensional.

Finally, if $x \in X$ then there exists $\widetilde{x} \in B$ such that $h(\widetilde{x}) = x$. Hence

$$d(x, \widetilde{x}) = d(h(\widetilde{x}), \widetilde{x}) \leq \epsilon^*, \tag{11.24}$$

proving (11.13). □

REMARK 11.1. While we needed g to be Anosov in order to construct h, the heart of the remaining results is contained in the observation that if h maps f to g, g is expansive and x is an equicontinuity point for f and f^{-1} then h is constant on some neighborhood of x.

CHAPTER 12

Generic Mappings on Manifolds

Overview: In this final section we sketch the corresponding results for mappings. In some respects these are easier because the p.l. maps (or smooth maps) are dense in $C(X;X)$ for X a p.l. manifold (resp. a smooth manifold).

With $C(X_1;X_2)$ denoting the space of continuous maps equipped with the sup metric, let $C_I(X_1;X_2)$ and $C_S(X_1;X_2)$ denote the subsets of injective and surjective maps, respectively.

PROPOSITION 12.1. *$C_S(X_1;X_2)$ is a closed subset of $C(X_1;X_2)$ and $C_I(X_1;X_2)$ is a G_δ subset.*

PROOF. For any open subset U of X_2, $\{f : f(X_1) \subset U\}$ is open in $C(X_1;X_2)$. Taking the union over all proper open subsets of X_2 we obtain the complement of $C_S(X_1;X_2)$. So C_S is closed. As for C_I, let $\Delta_j \subset X_j \times X_j$ denote the diagonal, and for $\epsilon > 0$ let $Z_\epsilon \subset X_1 \times X_1$ denote the complement of the ϵ-neighborhood of Δ_1. For each $\epsilon > 0$ $\{f : (f \times f)(Z_\epsilon) \cap \Delta_2 = \emptyset \}$ is open. Intersecting over positive rational ϵ's we obtain $C_I(X_1;X_2)$. So C_I is a G_δ. □

A continuous map $f : X_1 \to X_2$ is called *light* if $f^{-1}(y)$ is either empty or zero-dimensional for all $y \in X_2$ (see Hocking and Young, 1961). Let $C_L(X_1;X_2)$ denote the subset of light maps. Of course, $C_I \subset C_L$.

A continuous f is light iff for C a closed, connected subset of X_1, $f(C)$ a singleton subset of X_2 implies C is a singleton subset of X_1. The closed, connected subsets of a space X form a closed subset of $C(X)$. In fact, this subset is exactly $CT(1_X)$ (see the discussion surrounding ((1.26))).

LEMMA 12.2. (a) *If $f : X_1 \to X_2$ is light and A is a zero-dimensional subset of X_2 then $f^{-1}(A)$ is either empty or zero-dimensional.*

(b) *If $f : X_1 \to X_2$ and $g : X_2 \to X_3$ are light then $g \circ f : X_1 \to X_3$ is light.*

(c) *$C_L(X_1;X_2)$ is a G_δ subset of $C(X_1;X_2)$.*

(d) *If X is homeomorphic to a p.l. manifold then $C_L(X;X)$ is a residual subset of $C(X;X)$.*

PROOF. (a): Let C be a component of $f^{-1}(A)$. As $f(C)$ is a connected subset of A and A is zero-dimensional, $f(C)$ is a singleton set. Because f is light. C is a singleton. Thus, $f^{-1}(A)$ is zero-dimensional or empty.

(b): $(g \circ f)^{-1}(y) = f^{-1}(g^{-1}(y))$. Apply (a).

(c): For each $\epsilon > 0$, the set of $C \in CT(1_{X_1})$ such that diameter $C \geq \epsilon$ is a closed subset of $C(X_1)$. So, $\{f \in C(X_1;X_2) :$ diameter $f(C) > 0$ for all $C \in CT(1_{X_1})$ with diameter $\geq \epsilon\}$ is an open set for each positive ϵ. Intersecting over positive rational ϵ's we obtain $C_L(X_1;X_2)$. So C_L is a G_δ.

(d): First use p.l. approximation to get a p.l. map f_1 close to f and then use general position for maps, Rourke and Sanderson (1972) Theorem 5.4, to get a light map close to f_1. □

In adapting our earlier proofs to replace the assumption $f \in H(X)$ by $f \in C_L(X;X)$, we frequently use:

LEMMA 12.3. *Let $K_1, \dots, K_k$ be perfect, locally connected subsets of X_1. Let $x_i \in K_i$ for $i = 1, \dots, k$ and $\epsilon > 0$. If $f : X_1 \to X_2$ is light then there are points $\widetilde{x}_i \in K_i$ for $i = 1, \dots, k$ such that $d(x_i, \widetilde{x}_i) < \epsilon$, $\widetilde{x}_i \neq \widetilde{x}_j$ and $f(\widetilde{x}_i) \neq f(\widetilde{x}_j)$ for $i, j = 1, \dots, k$ and $j \neq i$.*

PROOF. For $k = 1$ no move is necessary. Use induction on k. By the inductive hypothesis we can assume that $\{x_1, \dots, x_{k-1}\}$ and $\{f(x_1), \dots, f(x_{k-1})\}$ are lists of $k-1$ distinct points. Because f is light, the set $\cup_{i=1}^{k-1} f^{-1}(f(x_i))$ is zero-dimensional. Since K_k is perfect and locally connected the x_k component of $V_\epsilon(x_k) \cap K_k$ contains a point $\widetilde{x}_k$ not in the set. □

For the set of continuous maps $C(X;X)$ and an $H(X)$ invariant basis $\mathcal{U}$ we can define subsets $G^1(K, K_1, \epsilon, m, \mathcal{U})$, $G^3(K, \epsilon, m, \mathcal{U})$, $G^4(K, \epsilon)$ and $G^8(\epsilon)$ subsets of $C(X;X)$ exactly as before using (6.1), (6.11), (6.27) and (11.1) respectively. With $\mathcal{U}_B$ the base of balls in X, we define the subset $C^*_{man}(X)$ of $C(X;X)$ by

$$C^*_{man}(X) = \{f \in C_L(X;X) : f \text{ satisfies } (12.2)\} \tag{12.1}$$

$$\left\{ \begin{array}{c} \text{for all nonempty closed subsets } K,\ K_1 \text{ of } X, \\ \text{all } \epsilon > 0 \text{ and all positive integers } m, \\ f \in G^1(K, K_1, \epsilon, m, \mathcal{U}_B) \cap G^3(K, \epsilon, m, \mathcal{U}) \cap G^4(K, \epsilon) \cap G^8(\epsilon) \\ \text{and } \mathcal{P}_m(f) = \mathcal{C}_m(f). \end{array} \right. \tag{12.2}$$

THEOREM 12.4. *If X is homeomorphic to a piecewise linear manifold of dimension at least 2, with $\partial X = \emptyset$ if dimension $X = 2$, then $C^*_{man}(X;X)$ is a residual subset of $C(X;X)$.*

A mapping f in $C^*_{man}(X;X)$ satisfies the following conditions:

(1) *f is a light map.*

(2) *Let K be a closed, forward invariant subset of X with $x \in K$. For every $\epsilon > 0$, $V_\epsilon(K)$ contains an attractor A such that the proper basin $W^s_f(A) \backslash A$ meets $V_\epsilon(K)$.*

(3) *Let K be a chain transitive subset for f. For every $\epsilon > 0$ and positive integer m there is an inward set U for f with an (ϵ, k)-periodic decomposition $\{U_0, \dots, U_{k-1}\}$ of type $\mathcal{U}_B$ such that $k \geq m$, $d(K, U) < \epsilon$ and U_0 contains a topological horseshoe for f^k.*

(4) *$\Omega(f) = \mathcal{C}(f)$ is a Cantor set in which $\Pi_{\infty,\infty}(f, \mathcal{U}_B)$ is a residual subset. The periodic points* Per(f) *are dense in $\mathcal{C}(f)$. $W^s_f(\Pi_{\infty,\infty}(f, \mathcal{U}_B))$ is a residual subset of X.*

(5) *A chain component B is terminal iff $B \subset \Pi_\infty(f)$. The terminal chain components form a residual subset of the Cantor space $\mathcal{B}_f$. The $\Pi_{\infty,\infty}(f, \mathcal{U}_B)$ chain components form a residual subset of $CT(f)$ which is disjoint from the set of periodic orbits. The latter is also dense in $CT(f)$.*

(6) *The set of points at which f is chain continuous is the residual subset of X:*

$$W_f^s(\Pi_\infty(f)) = \{x : \omega(x,f) = \omega\mathcal{C}(x,f)\}.$$

In particular, f is an almost equicontinuous map.

(7) *If R is a repellor for f then R contains attractors, has a nonempty interior and satisfies*

$$R = \overline{\cup\{W_f^s(A_1)\backslash A_1 : A_1 \subset R \text{ and } A_1 \text{ is an attractor}\}}. \tag{12.3}$$

Each repellor contains a countable infinity of disjoint attractors.

(8) *If A is an attractor then $\mathcal{C}(\partial A, f) = \partial A$ and so ∂A is a quasi-attractor.*

PROOF. Each G^1, G^3, G^4 and G^8 is open just as before as in the replacement of the class of subsets by a countable class. Density is obtained by adapting the proofs of the crushing arguments. The hypotheses on X show that either $\partial X = \emptyset$ or else the dimension of ∂X is at least 2. Hence, X is generalized homogeneous on $X\backslash\partial X$ and X is strictly spongy. The sponges can be chosen to be arcs in $X\backslash\partial X$ or ∂X. As such we can avoid not only any finite set but any zero-dimensional set. We leave as exercises the adjustments needed to extend Propositions 4.3, 4.9, 4.10, 4.11 and 5.4 to maps. The analogue of Proposition 10.2 is much easier for maps because we can directly use the p.l. approximation theorem. We are no longer trying to approximate a homeomorphism by a p.l. homeomorphism. Hence, Proposition 10.3 follows from maps without the extra condition that $n \neq 3, 4$ which $\partial X \neq \emptyset$.

Conditions (1), (2), (3) follow from the definitions when $f \in C^*_{man}$. The equality of $\Omega(f)$ and $\mathcal{C}(f)$ follows from Proposition 3.5. Follow Theorem 6.4 to prove that $W_f^s(\Pi_{\infty,\infty}(f,\mathcal{U}_B))$ is residual in X. Then density of periodic orbits and $\Pi_{\infty,\infty}(f,\mathcal{U}_B)$ chain components in $CT(f)$ follows from (3). In particular, every chain component can be approximated by periodic orbits and by $\Pi_{\infty,\infty}(f,\mathcal{U}_B)$ chain components. So the disjoint sets $\text{Per}(f)$ and $\Pi_{\infty,\infty}(f,\mathcal{U}_B)$ are each dense in $\mathcal{C}(f)$, and $\mathcal{C}(f)$ has no isolated points. Similarly, the $\Pi_{\infty,\infty}(f,\mathcal{U}_B)$ chain components are dense in $\mathcal{B}_f$ and the latter has no isolated points. A chain component is terminal iff it is in $\Pi_\infty(f)$ by Proposition 2.3. For condition (6) follow Proposition 11.5. For conditions (7) and (8) follow the corresponding portions of Theorem 6.3 and Theorem 6.9. □

Bibliography

[1] E. Akin (1969) "Manifold phenomena in the theory of polyhedra", *Trans. Amer. Math. Soc.* **142**, 413-473.

[2] ——— (1993) *The general topology of dynamical systems*, Amer. Math. Soc., Providence.

[3] ——— (1996) "On chain continuity", *Discrete and Cont. Dynam. Sys.* **2**, 111-120.

[4] ——— (1999) Simplicial dynamical systems, *Memoirs AMS* No. 667, Amer. Math. Soc., Providence.

[5] ——— (2000) "Stretching the Oxtoby-Ulam theorem", *Colloq. Math* **84/85**, 83-94.

[6] E. Akin, J. Auslander and K. Berg (1996) "When is a transitive map chaotic?", pp. 25-40 in *Conference in ergodic theory and probability* (V. Bergelson, K. March and J. Rosenblatt, eds.), Walter de Gruyter, Berlin.

[7] S. Alpern and V.S. Prasad (2000) *Typical properties of volume preserving homeomorphisms,* Cambridge Tracts in Math., 139, Cambridge Univ. Press, Cambridge.

[8] N. Aoki and K. Hiraide (1994) *Topological theory of dynamical systems,* Elsevier, Amsterdam.

[9] J. Auslander (1988) *Minimal flows and their extensions*, Elsevier, Amsterdam.

[10] J. Auslander and J. Yorke (1980) "Interval maps, factors of maps and chaos", *Tohoku Math. J.* **32**, 177-188.

[11] N. Bhatia and G. Szegö (1967) *Dynamical systems: stability theory and applications*, Lect. Notes in Math. No. 35, Springer-Verlag, Berlin.

[12] N. Bourbaki (1966) *Elements of mathematics: general topology, part 2*, Addison-Wesley, Reading.

[13] J. Buescu and I. Stewart (1995) "Lyapunov stability and adding machines", *Ergod. Th. and Dynam.* Sys. **15**, 271-290.

[14] M. Cohen (1969) "A general theory of relative regular neighborhoods", *Trans. Amer. Math. Soc.* **136**, 189-230.

[15] C. Conley (1978) *Isolated invariant sets and the Morse index*, CBMS Reg. Conf. Ser. in Math No. 38, Amer. Math. Soc., Providence.

[16] J. Dancis (1976) "General position maps for topological manifolds in the 2/3rds range", *Trans. Amer. Math. Soc.* **216**, 249-266.

[17] M. Denker, C. Grillenberger and K. Sigmund (1976) *Ergodic theory on compact spaces*, Lect. Notes in Math. No. 527, Springer-Verlag, Berlin.

[18] H. Furstenberg (1981) *Recurrence in ergodic theory and combinatorial number theory*, Princeton Univ. Press, Princeton.

[19] E. Glasner and J. King (1998) "A zero-one law for dynamical properties", pp. 213-242 in *Topological dynamics and applications: a volume in honor of Robert Ellis,* (M. Nerurkar, D. Dokken and D. Ellis, eds.), Contemp. Math. No. 215, Amer. Math. Soc., Providence.

[20] E. Glasner and B. Weiss (1993) "Sensitive dependence on initial condition", *Nonlinearity* **6**, 1067-1075.

[21] ——— (2001) "The topological Rohlin property and topological entropy", *Am. J. Math.* **123**, 1055-1070.

[22] W. Gottschalk and G. Hedlund (1955) *Topological dynamics,* Amer. Math. Soc., Providence.

[23] J. Guckenheimer and P. Holmes (1983) *Nonlinear oscillations, dynamical systems, and bifurcations of vector fields,* Springer-Verlag, New York.

[24] M. Herman (1977) "Mesure de Lebesgue et nombre de rotation", pp. 271-293 in *Geometry and topology,* Lect. Notes in Math. No. 597, Springer-Verlag, Berlin.

[25] M. Hirsch and M. Hurley (1997) "Connected components of attractors and other stable sets", *Aequ. Math.* **53**, 308- 323.

[26] J. Hocking and G. Young (1961) *Topology,* Addison-Wesley, Reading, Mass.

[27] J. Hudson (1969) *Piecewise linear topology*, W. A. Benjamin, New York.

[28] M. Hurley (1995) "Generic homeomorphisms have no smallest attractor", *Proc. Amer. Math. Soc.* **123**, 1277-1280.

[29] ——— (1996a) "Properties of attractors of generic homeomorphisms", *Ergod. Th. and Dynam. Sys.* **16**, 1297-1310.

[30] ——— (1996b) "On proofs of the C^0 density theorem", *Proc. Amer. Math. Soc.* **124**, 1305-1309.

[31] M. Jakobson (1980) "Construction of invariant measures absolutely continuous with respect to dx for some maps of the interval", pp. 246-257 in *Global theory of dynamical systems*, Lect. Notes in Math. No. 819, Springer-Verlag, Berlin.

[32] A. Katok and B. Hasselblatt (1995) *Introduction to the modern theory of dynamical systems*, Cambridge U. Press, Cambridge.

[33] J. L. Kelley (1955) *General topology*, Van Nostrand, New York.

[34] J. Kennedy (1996) "The topology of attractors", *Ergod. Th. and Dynam. Sys.* **16**, 1311-1322.

[35] K. Kuratowski (1932) "Les fonctions semi-continues dans l'espace des ensembles fermées", *Fund. Math.* **18**, 148- 159.

[36] I. Melbourne, M. Dellnitz and M. Golubitsky (1993) "The structure of symmetric attractors", *Arch. Rat. Mech. Anal.* **123**, 75-99.

[37] P. Michor (1980) *Manifolds of differentiable mappings*, Shiva, Kent.

[38] J. Milnor (1963) *Morse theory*, Princeton U. Press, Princeton.

[39] ——— (1985) "On the concept of attractor", *Commun. Math. Phys.* **99**, 177-195; "Correction and remarks", *Commun. Math. Phys.* **102**, 517-519.

[40] E. Moise (1952) "Affine structures in three manifolds: V. the triangulation theorem and hauptvermutung", *Annals of Math.* **56**, 96-114.

[41] J. Munkres (1963) *Elementary differential topology*, Princeton U. Press, Princeton.

[42] Z. Nitecki (1971) *Differentiable dynamics: an introduction to the orbit structure of diffeomorphisms*, MIT U. Press, Cambridge.

[43] Z. Nitecki and M. Shub (1975) "Filtrations, decompositions and explosions", *Amer. J. Math.* **107**, 1029.

[44] A. V. Ostrovsky (2000) "Stable maps of Polish spaces", *Proc. A.M.S.* **128**, 3081-3089.

[45] J. Oxtoby (1977) "Diameters of arcs and the gerrymandering problem", *Am. Math. Monthly* **84**, 155-162.

[46] ——— (1980) *Measure and category*, Springer-Verlag, New York.

[47] J. Oxtoby and S. Ulam (1941) "Measure-preserving homeomorphisms and metrical transitivity", *Ann. of Math.* **42**, 874-920.

[48] J. Palis, C. Pugh, M. Shub and D. Sullivan (1975) "Genericity theorems in topological dynamics," pp. 241-250 in *Dynamical systems – Warwick 1974*, Lect. Notes in Math No. 468, Springer-Verlag, Berlin.

[49] S. Pilyugin (1994) *The space of dynamical systems with the C^0-topology*, Lect. Notes in Math No. 1571, Springer-Verlag, Berlin.

[50] ——— (1999) *Shadowing in dynamical systems*, Lect. Notes in Math No. 1706, Springer-Verlag, Berlin.

[51] S. Pilyugin and O. Plamenevskaya (1999) "Shadowing is generic", *Topology Appl.* **97**, 253-266.

[52] C. Robinson, (1995) *Dynamical systems*, CRC Press, Boca Raton.

[53] C. Rourke and B. Sanderson (1972) *Introduction to piecewise linear topology*, Springer-Verlag, Berlin.

[54] M. Sears (1971) "Topologies on the set of self homeomorphisms of the Cantor set", *Mat. Casopis. Sloven. Akad. Vied* **21**, 227-232.

[55] ——— (1972) "Expansive self homeomorphisms of the Cantor set", *Math. Systems Theory* **6**, 129-132.

[56] M. Shub (1972) "Structurally stable diffeomorphisms are dense", *Bull. Amer. Math. Soc.* **78**, 817-818.

[57] W. Sierpinski (1930) "Sur une propriété des ensembles G_δ", *Fund. Math.* 173-180.

[58] ——— (1956) *General topology* (trans. C.C. Krieger), University of Toronto Press, Toronto.

[59] J. Stallings (1968) *Lectures on polyhedral topology*, de Gruyter, Berlin.

[60] T. tom Dieck (1987) *Transformation groups*, Tata Institute of Fundamental Research, Bombay.

[61] F. Takens (1971) "On Zeeman's tolerance stability conjecture", pp. 209-219 in *Manifolds-Amsterdam 1970*, Lect. Notes in Math. No. 197, Springer-Verlag, Berlin.

[62] P. Walters (1978) "On the pseudo orbit tracing property and its relationship to stability", pp. 231-244 in *The structure of attractors in dynamical systems*, Lect. Notes in Math. No. 668, Springer-Verlag, Berlin.

Index

Editorial Information

To be published in the *Memoirs*, a paper must be correct, new, nontrivial, and significant. Further, it must be well written and of interest to a substantial number of mathematicians. Piecemeal results, such as an inconclusive step toward an unproved major theorem or a minor variation on a known result, are in general not acceptable for publication. Papers appearing in *Memoirs* are generally longer than those appearing in *Transactions*, which shares the same editorial committee.

As of April 1, 2003, the backlog for this journal was approximately 4 volumes. This estimate is the result of dividing the number of manuscripts for this journal in the Providence office that have not yet gone to the printer on the above date by the average number of monographs per volume over the previous twelve months, reduced by the number of volumes published in four months (the time necessary for preparing a volume for the printer). (There are 6 volumes per year, each containing at least 4 numbers.)

A Consent to Publish and Copyright Agreement is required before a paper will be published in the *Memoirs*. After a paper is accepted for publication, the Providence office will send a Consent to Publish and Copyright Agreement to all authors of the paper. By submitting a paper to the *Memoirs*, authors certify that the results have not been submitted to nor are they under consideration for publication by another journal, conference proceedings, or similar publication.

Information for Authors

Memoirs are printed from camera copy fully prepared by the author. This means that the finished book will look exactly like the copy submitted.

The paper must contain a *descriptive title* and an *abstract* that summarizes the article in language suitable for workers in the general field (algebra, analysis, etc.). The *descriptive title* should be short, but informative; useless or vague phrases such as "some remarks about" or "concerning" should be avoided. The *abstract* should be at least one complete sentence, and at most 300 words. Included with the footnotes to the paper should be the 2000 *Mathematics Subject Classification* representing the primary and secondary subjects of the article. The classifications are accessible from `www.ams.org/msc/`. The list of classifications is also available in print starting with the 1999 annual index of *Mathematical Reviews*. The Mathematics Subject Classification footnote may be followed by a list of *key words and phrases* describing the subject matter of the article and taken from it. Journal abbreviations used in bibliographies are listed in the latest *Mathematical Reviews* annual index. The series abbreviations are also accessible from `www.ams.org/publications/`. To help in preparing and verifying references, the AMS offers MR Lookup, a Reference Tool for Linking, at `www.ams.org/mrlookup/`. When the manuscript is submitted, authors should supply the editor with electronic addresses if available. These will be printed after the postal address at the end of the article.

Electronically prepared manuscripts. The AMS encourages electronically prepared manuscripts, with a strong preference for $\mathcal{A}\mathcal{M}\mathcal{S}$-LaTeX. To this end, the Society has prepared $\mathcal{A}\mathcal{M}\mathcal{S}$-LaTeX author packages for each AMS publication. Author packages include instructions for preparing electronic manuscripts, the *AMS Author Handbook*, samples, and a style file that generates the particular design specifications of that publication series. Though $\mathcal{A}\mathcal{M}\mathcal{S}$-LaTeX is the highly preferred format of TeX, author packages are also available in $\mathcal{A}\mathcal{M}\mathcal{S}$-TeX.

Authors may retrieve an author package from e-MATH starting from `www.ams.org/tex/` or via FTP to `ftp.ams.org` (login as `anonymous`, enter username as password, and type `cd pub/author-info`). The *AMS Author Handbook* and the *Instruction Manual* are available in PDF format following the author packages link from `www.ams.org/tex/`. The author package can be obtained free of charge by sending email to `pub@ams.org` (Internet) or from the Publication Division, American Mathematical Society, 201 Charles St., Providence, RI 02904, USA. When requesting an author package, please specify $\mathcal{A}\mathcal{M}\mathcal{S}$-LaTeX or $\mathcal{A}\mathcal{M}\mathcal{S}$-TeX, Macintosh or IBM (3.5) format, and the publication in which your paper will appear. Please be sure to include your complete mailing address.

Sending electronic files. After acceptance, the source file(s) should be sent to the Providence office (this includes any TeX source file, any graphics files, and the DVI or PostScript file).

Before sending the source file, be sure you have proofread your paper carefully. The files you send must be the EXACT files used to generate the proof copy that was accepted for publication. For all publications, authors are required to send a printed copy of their paper, which exactly matches the copy approved for publication, along with any graphics that will appear in the paper.

TeX files may be submitted by email, FTP, or on diskette. The DVI file(s) and PostScript files should be submitted only by FTP or on diskette unless they are encoded properly to submit through email. (DVI files are binary and PostScript files tend to be very large.)

Electronically prepared manuscripts can be sent via email to `pub-submit@ams.org` (Internet). The subject line of the message should include the publication code to identify it as a Memoir. TeX source files, DVI files, and PostScript files can be transferred over the Internet by FTP to the Internet node `e-math.ams.org` (130.44.1.100).

Electronic graphics. Comprehensive instructions on preparing graphics are available at `www.ams.org/jourhtml/graphics.html`. A few of the major requirements are given here.

Submit files for graphics as EPS (Encapsulated PostScript) files. This includes graphics originated via a graphics application as well as scanned photographs or other computer-generated images. If this is not possible, TIFF files are acceptable as long as they can be opened in Adobe Photoshop or Illustrator. No matter what method was used to produce the graphic, it is necessary to provide a paper copy to the AMS.

Authors using graphics packages for the creation of electronic art should also avoid the use of any lines thinner than 0.5 points in width. Many graphics packages allow the user to specify a "hairline" for a very thin line. Hairlines often look acceptable when proofed on a typical laser printer. However, when produced on a high-resolution laser imagesetter, hairlines become nearly invisible and will be lost entirely in the final printing process.

Screens should be set to values between 15% and 85%. Screens which fall outside of this range are too light or too dark to print correctly. Variations of screens within a graphic should be no less than 10%.

Inquiries. Any inquiries concerning a paper that has been accepted for publication should be sent directly to the Electronic Prepress Department, American Mathematical Society, 201 Charles St., Providence, RI 02904, USA.

Editors

This journal is designed particularly for long research papers, normally at least 80 pages in length, and groups of cognate papers in pure and applied mathematics. Papers intended for publication in the *Memoirs* should be addressed to one of the following editors. In principle the Memoirs welcomes electronic submissions, and some of the editors, those whose names appear below with an asterisk (*), have indicated that they prefer them. However, editors reserve the right to request hard copies after papers have been submitted electronically. Authors are advised to make preliminary email inquiries to editors about whether they are likely to be able to handle submissions in a particular electronic form.

Algebra to KAREN E. SMITH, Department of Mathematics, University of Michigan, 525 University, Suite 2832, Ann Arbor, MI 48109-1109; email: kesmith@lsa.umich.edu

Algebraic geometry and commutative algebra to LAWRENCE EIN, Department of Mathematics, University of Illinois, 851 S. Morgan (M/C 249), Chicago, IL 60607-7045; email: ein@uic.edu

Algebraic topology and cohomology of groups to STEWART PRIDDY, Department of Mathematics, Northwestern University, 2033 Sheridan Road, Evanston, IL 60208-2730; email: priddy@math.nwu.edu

Combinatorics and Lie theory to SERGEY FOMIN, Department of Mathematics, University of Michigan, Ann Arbor, Michigan 48109-1109; email: fomin@umich.edu

Complex analysis and complex geometry to DUONG H. PHONG, Department of Mathematics, Columbia University, 2990 Broadway, New York, NY 10027-0029; email: phong@math.columbia.edu

***Differential geometry and global analysis** to LISA C. JEFFREY, Department of Mathematics, University of Toronto, 100 St. George St., Toronto, ON Canada M5S 3G3; email: jeffrey@math.toronto.edu

Dynamical systems and ergodic theory to ROBERT F. WILLIAMS, Department of Mathematics, University of Texas, Austin, Texas 78712-1082; email: bob@math.utexas.edu

Functional analysis and operator algebras to DAN VOICULESCU, Department of Mathematics, University of California, Berkeley, 970 Evans Hall, Floor 9, Berkeley, CA 94720-0001; email: dvv@math.berkeley.edu

Geometric topology, knot theory and hyperbolic geometry to ABIGAIL A. THOMPSON, Department of Mathematics, University of California, Davis, Davis, CA 95616-5224; email: thompson@math.ucdavis.edu

Harmonic analysis to ALEXANDER NAGEL, Department of Mathematics, University of Wisconsin, 480 Lincoln Drive, Madison, WI 53706-1313; email: nagel@math.wisc.edu

Harmonic analysis, representation theory, and Lie theory to ROBERT J. STANTON, Department of Mathematics, The Ohio State University, 231 West 18th Avenue, Columbus, OH 43210-1174; email: stanton@math.ohio-state.edu

***Logic** to THEODORE SLAMAN, Department of Mathematics, University of California, Berkeley, CA 94720-3840; email: slaman@math.berkeley.edu

Number theory to HAROLD G. DIAMOND, Department of Mathematics, University of Illinois, 1409 W. Green St., Urbana, IL 61801-2917; email: diamond@math.uiuc.edu

***Ordinary differential equations, and applied mathematics** to PETER W. BATES, Department of Mathematics, Michigan State University, East Lansing, MI 48824-1027; email: peter@math.msu.edu

***Partial differential equations** to PATRICIA E. BAUMAN, Department of Mathematics, Purdue University, West Lafayette, IN 47907-1395' email: bauman@math.purdue.edu

***Probability and statistics** to KRZYSZTOF BURDZY, Department of Mathematics, University of Washington, Box 354350, Seattle, Washington 98195-4350; email: burdzy@math.washington.edu

***Real analysis and partial differential equations** to DANIEL TATARU, Department of Mathematics, University of California, Berkeley, Berkeley, CA 94720; email: tataru@math.berkeley.edu

All other communications to the editors should be addressed to the Managing Editor, WILLIAM BECKNER, Department of Mathematics, University of Texas, Austin, TX 78712-1082; email: beckner@math.utexas.edu.

Titles in This Series

783 **Ethan Akin, Mike Hurley, and Judy A. Kennedy,** Dynamics of topologically generic homeomorphisms, 2003

782 **Masaaki Furusawa and Joseph A. Shalika,** On central critical values of the degree four L-functions for GSp(4): The Fundamental Lemma, 2003

781 **Marcin Bownik,** Anisotropic Hardy spaces and wavelets, 2003

780 **S. Marmi and D. Sauzin,** Quasianalytic monogenic solutions of a cohomological equation, 2003

779 **Hansjörg Geiges,** h-principles and flexibility in geometry, 2003

778 **David B. Massey,** Numerical control over complex analytic singularities, 2003

777 **Robert Lauter,** Pseudodifferential analysis on conformally compact spaces, 2003

776 **U. Haagerup, H. P. Rosenthal, and F. A. Sukochev,** Banach embedding properties of non-commutative L^p-spaces, 2003

775 **P. Lochak, J.-P. Marco, and D. Sauzin,** On the splitting of invariant manifolds in multidimensional near-integrable Hamiltonian systems, 2003

774 **Kai A. Behrend,** Derived ℓ-adic categories for algebraic stacks, 2003

773 **Robert M. Guralnick, Peter Müller, and Jan Saxl,** The rational function analogue of a question of Schur and exceptionality of permutation representations, 2003

772 **Katrina Barron,** The moduli space of $N = 1$ superspheres with tubes and the sewing operation, 2003

771 **Shigenori Matsumoto,** Affine flows on 3-manifolds, 2003

770 **W. N. Everitt and L. Markus,** Elliptic partial differential operators and symplectic algebra, 2003

769 **Jie Wu,** Homotopy theory of the suspensions of the projective plane, 2003

768 **R. Höpfner and E. Löcherbach,** Limit theorems for null recurrent Markov processes, 2003

767 **Po Hu,** S-modules in the category of schemes, 2003

766 **Su Gao and Alexander S. Kechris,** On the classification of Polish metric spaces up to isometry, 2003

765 **Robert Bieri and Ross Geoghegan,** Connectivity properties of group actions on non-positively curved spaces, 2003

764 **J. Spandaw,** Noether-Lefschetz problems for degeneracy loci, 2003

763 **Yasuyuki Kachi and Eiichi Sato,** Segre's reflexivity and an inductive characterization os hyperquadrics, 2002

762 **Leiba Rodman, Ilya M. Spitkovsky, and Hugo Woerdeman,** Abstract band method via factorization, positive and band extensions of multivariable almost periodic matrix functions, and spectral estimation, 2002

761 **Oliver Druet and Emmanuel Hebey,** The AB program in geometric analysis : Sharp Sobolev inequalities and related problems, 2002

760 **Markus Banagl,** Extending intersection homology type invarients to non-Witt spaces, 2002

759 **Donald M. Davis,** From representation theory to homotopy groups, 2002

758 **Alan Forrest, John Hunton, and Johannes Kellendonk,** Topological invariants for projection method patterns, 2002

757 **Douglas Bowman,** q-difference operators, orthogonal polynomials, and symmetric expansions, 2002

756 **José Ignacio Cogolludo-Agustín,** Topological invariants of the complement to arrangements of rational plane curves, 2002

755 **M. A. Mandell and J. P. May,** Equivariant orthogonal spectra and S-modules, 2002

TITLES IN THIS SERIES